THINKING GEOMETRICALLY

Digital Formations

Steve Jones
General Editor

Vol. 2

PETER LANG
New York • Washington, D.C./Baltimore • Bern
Frankfurt am Main • Berlin • Brussels • Vienna • Oxford

John T. Waisanen

THINKING GEOMETRICALLY

Re-Visioning Space for a Multimodal World

EDITED AND PREFACE BY
Jennifer Daryl Slack

POSTSCRIPT BY
Kevin G. Barnhurst

PETER LANG
New York • Washington, D.C./Baltimore • Bern
Frankfurt am Main • Berlin • Brussels • Vienna • Oxford

Library of Congress Cataloging-in-Publication Data

Waisanen, John T.
Thinking geometrically: re-visioning space for a multimodal world /
John T. Waisanen; edited by Jennifer Daryl Slack.
p. cm. — (Digital formations; vol. 2)
Includes bibliographical references.
1. Engineering. 2. Creative thinking.
I. Slack, Jennifer Daryl. II. Title.
TA145 .W35 620—dc21 2001038978
ISBN 0-8204-5184-3
ISSN 1526-3169

Die Deutsche Bibliothek-CIP-Einheitsaufnahme

Waisanen, John T.:
Thinking geometrically: re-visioning space for a multimodal world /
John T. Waisanen: Ed. by: Jennifer Daryl Slack.
–New York; Washington, D.C./Baltimore; Bern;
Frankfurt am Main; Berlin; Brussels; Vienna; Oxford: Lang.
(Digital formations; Vol. 2)
ISBN 0-8204-5184-3

Cover art: detail from *Two Horizons* by jd slack

Cover design by Joni Holst

For the Waisanen family:

Frederick, Helen, Susan, John, Ellen, and Thomas

Contents

AUTHOR'S ACKNOWLEDGMENTS

I would like to thank the Michigan Technological University Graduate School for granting a three-year doctoral fellowship to start this project and subsequently granting a one-term fellowship to conclude it. This support was essential.

I also wish to thank the advisors who have guided me in the course of this work. The constructive criticism and encouragement that Jennifer Daryl Slack, Barry Pegg, Walter Olson, and Diane Shoos have shared with me have been invaluable throughout my research. Especially, I would like to thank Jennifer Daryl Slack and Marjorie Lindley for their extracurricular (and extraordinary) commitment to review and correct an earlier version of this work.

I deeply appreciate the support of advisors Eric Freedman and Cindy Selfe during the early years of my studies. Discussions with fellow graduate students, especially Margaret Faler Sweany, have also helped me to develop a deeper awareness of the relation between contemporary technology and critical thought.

Without the unflagging support and sacrifice of my family—especially Susan, but also my children John, Ellen, and Thomas, and my parents Frederick and Helen—pursuing this project to its completion would have been impossible.

Equally essential in this regard was the encouragement, scheduling flexibility, and financial stability provided by Edmond Progar and Richard Smelker, and Virtual Services, Inc., where I engaged in contract writing, course development, and training in the interim years between fellowships.

Many earlier influences are foundational to this work. Dan Zink's and Harold Philpot's course in automotive design drawing provided me with the necessary fundamentals of descriptive geometry. John Mills (Wayne State University) encouraged my interest in industrial design methodology and freehand drawing. Joseph Woelfel (then of Michigan State University) inspired my earliest interests in visualizing spatial patterns of information.

Over the years, many other colleagues and friends have served, though less conspicuously, as resources for this project. Whether anybody knew (or knows) of this contribution or not, I would also like to say thanks. Special thanks are due especially to Rob Reynolds and Audrey Arntzen. Had our paths never crossed, I doubt that any of this would have come about.

An earlier, and abridged, version of several parts of the first chapter was presented as "Thinking Geometrically" at the May 1996 conference of the International Communication Association. I wish to thank the reviewers and participants in that conference for their interest and encouragement.

John T. Waisanen

EDITOR'S ACKNOWLEDGMENTS

I would like to express my gratitude for the support I received that made it possible to give John Waisanen's work the exposure it deserves. I am deeply grateful to the following: Sung Lee, who as Dean of the Michigan Technological University Graduate School, granted John the fellowship that allowed him to finish writing; Marilyn Urion, Assistant to the Dean, who took on the duties of Dean after Lee's departure and found summer support money for the two graduate assistants who helped me with the final stages of manuscript editing and preparation; Gordon Coonfield and Cynthia Weber, the most conscientious of assistants, for taking on their work with enthusiasm and intellectual rigor; Wellesley Pereira, for assistance with clarifying some aspects of descriptive geometry; the Michigan Technological University Graduate School, for a scholarship to assist in the final phase of production; the Michigan Technological University Department of Humanities, especially Marjorie Lindley for early typing, corrections, and computer management; and Robert Johnson, who, as department chair, created an environment within which it was possible to get the work done.

In addition, I would like to express a very special and heartfelt sense of appreciation to Susan Waisanen, John's wife, who has been simply extraordinary in supporting John's work both before and after his death.

Jennifer Daryl Slack

PREFACE

THINKING WITH THE TECHNICAL PEN

JENNIFER DARYL SLACK

Recently I went to my university bookstore in search of a technical pen. After admiring these fascinating (and relatively expensive) pens for years, I had decided to educate my hand in their use. After searching the store without success, I consulted the employee who handles drafting supplies. Where are the technical pens?

Although the employee did locate an inexpensive (and inferior) version of a technical pen, he told me that the bookstore will no longer be stocking technical pens, or for that matter, any traditional drafting supplies: technical pens, French curves, compass, protractor, and so on. What supplies remained were discounted 40% to move them off the shelves. Engineering students, he told me, are no longer being taught to draw using pen and pencil. Instead, their design training consists almost exclusively of working with computer software that does the drawing for them. His response underscores the pressing significance of John Waisanen's *Thinking Geometrically*.

I should not have been surprised by the huge discounts on drafting supplies. In lectures and books from the 1960s on, Mike Cooley, design engineer and trade union activist, has been predicting the demise of drawing and design skill as engineering, architecture, and design rely more and more on the computer for their execution. And Waisanen had for years been telling me that many engineers and designers he encountered did not know how to draw—often with embarrassing and sometimes even devastating consequences.

After the surprise subsided, alarm took its place. The education of engineers has the highest priority on my campus; very little is spared to ensure that engineering students have the best resources available to them. To the extent that engineering education at this campus no longer entails drawing, drawing is likely to be disappearing across the nation, just as Cooley predicted. For Cooley, this amounts to nothing less than the deskilling, or "proletarianization," of scientists and technologists; the introduction of Taylorism into intellectual work; and an overall suppression of the intellect (Cooley, 1980). Cooley acknowledges the enormous capabilities of computer-aided design (CAD), but he reminds us of the cultural tendency "to make value judgements about given technologies based on what they might achieve rather than what they have already achieved and are likely to continue to achieve within a given economic, political and social framework." The reliance on CAD, Cooley argues, entails "the subordination of the operator (designer) to the machine (computer), with the narrow specialization of Taylorism leading to the fragmentation of design skills and a loss of panoramic view of the design activity itself" (35).

In *Thinking Geometrically*, John Waisanen explores the sources, significance, and consequences of that "loss of panoramic view of the design activity itself." He also argues passionately for its cultivation. Ironically, contemporary life requires the cultivation of exactly the skill we are neglecting to value or master: that is, the skill of thinking geometrically. If Waisanen is right, and I am convinced that he is, the engineers we graduate today are ill-prepared to think—and design—in exactly those ways that contemporary circumstances most demand. With our fascination for new technologies, we have become seduced into believing that specialized technical (typically, computer) expertise equips us best for the times, when in practice the shift away from the cultivation of geometrical thinking virtually ensures a diminished ability to design—automobiles as well as life—effectively.

The need to design, however, is not unique to engineers, architects, planners, and those whose work we typically understand as engaged in "design." Rather, Waisanen argues, design is an activity in which we all engage: we design our lives, our everyday activities, our political and economic systems, our environment, our futures. To design effectively, Waisanen maintains, requires the ability to think geometrically. But where do we learn to design? Where and how is the capacity to think geometrically and design effectively cultivated? If engineers—whose job specifically requires the ability to design—are poorly equipped, where does that leave the rest of us? After thinking through Waisanen's argument, I can only conclude that we are becoming increasingly ill-equipped to think geometrically and design effectively as our education, training, and practice moves further away from focusing on the problem of understanding and manipulating the full range of complex relations that constitute contemporary life. Instead we are taught to specialize—to shy away from the difficult (and messy) work with the hand (such as working with the technical pen)—and to let the world as *already designed* dictate the territory for us. We learn to adjust, accommodate, adapt—not design. Every individual has something to learn from this book, then, about the need to think geometrically, the difficulties of doing so, and the consequences of not taking the challenge seriously. As designers of our lives, each of us—whether engineer, architect, professional designer or not—can benefit by learning to think geometrically.

Why Think Geometrically?

Waisanen's most straightforward articulation of what it means to think geometrically is "the ability to deal imaginatively with the multiple, that is, numerous and simultaneous, dimensions of complex and variable objects and relations in a structured, yet dynamic and interactive fashion. . . . [T]his means being able to construct mental models, to compare these models to real events, and to modify both the model and one's own behavior to arrive at a better fit between imaginative constructions and events in real space" (8–9). But like geometrical thinking itself, this definition must be understood as the condensation of multiple dimensions. To grasp its full significance, its meaning must be felt "on the body" so to speak. To facilitate that "sense" of thinking geometrically, let's look at individual parts of the definition, beginning with the phrase "multiple and simultaneous dimensions of complex and variable objects and relations."

Virtually everyone, from postmodern theorists and computer tycoons to spiritual guides, acknowledges that negotiating the terrain of everyday life has become a challenge of near overwhelming proportions. Theorists of the postmodern point to a variety of changes that mark the chaos of the contemporary: the collapse of metanarratives and cultural hierarchies; the shattering of modernist conceptions of space and time; the fragmentation of subject positions and identity; the challenge of multiculturalisms, new media environments, and globalization; the commercialization of nearly everything; and the proliferation of surfaces, images, and affects. Perhaps the world always was complex, but the attenuation of time and space render contemporary life a minefield to traverse and the stress levels palpable. In my favorite—if disturbing—statement on the nature of contemporary life, Thomas Merton (1989:86) captures the way our capacity for constructive action diminishes as we try valiantly to respond to the multiple and overwhelming demands placed on us:

> [T]here is a pervasive form of contemporary violence . . . activism and overwork. The rush and pressure of modern life are a form, perhaps the most common form, of its innate violence. To allow oneself to be carried away by a multitude of

> conflicting concerns, to surrender to too many demands, to commit oneself to too many projects, to want to help everyone in everything is to succumb to violence. More than that, it is cooperation in violence. The frenzy of the activist neutralizes his [or her] work . . . because it kills the root of inner wisdom which makes work fruitful.

The increasing popularity of Buddhism in the West highlights the need for maps within which to navigate this increasingly complex world. As Thich Nhat Hanh puts it, "We have so many compartments in our lives" (1991:34). It has become a daunting task to manage and coordinate all these individual compartments.

Representative of the dominant cultural tendency, Bill Gates (1999) approaches the problem from a limited business perspective that reduces the matter to the task of managing information. He argues that the speed and complexity of business is accelerating at such a rate that only the development of a "digital nervous system" can help us manage the evolving terrain:

> A digital nervous system is the corporate, digital equivalent of the human nervous system, providing a well-integrated flow of information to the right part of the organization at the right time. A digital nervous system consists of the digital processes that enable a company to perceive and react to its environment, to sense competitor challenges and customer needs, and to organize timely responses. A digital nervous system requires a combination of hardware and software; it's distinguished from a mere network of computers by the accuracy, immediacy, and richness of the information it brings to knowledge workers and the *insight* and *collaboration* made possible by the information. (xvii–xviii)

In Gates's formulation, the facility to gather, manage, and use information determines whether you win or lose. In the digital nervous system, Merton's sense of "inner wisdom" isn't even at issue.

Whether one's response is academic (such as theorizing the postmodern), spiritual (such as learning to meditate or pray), activist (such as developing a digital nervous system), or a combination of these, we are all forced to respond in some way, for this is the world in which we live. Engineers learn to save time by designing on computers; they develop specialities that allow them to focus—and thus maximize—their talents. Most of us struggle to continually "upgrade" our ability to navigate the multitude of conflicting concerns of everyday life. Our thirst for ways to "hold it all together" feeds an industry in the production of navigational aides: Franklin Planners and Palm Pilots to manage our time and memories, faster computers to get done whatever we do there faster, the Internet as a faster and more efficient source of commercial goods and information, intelligent agents to stand in for our thinking processes, and telephone answering services to stand in for us and keep track of demands on our time we might otherwise miss.

Education is the training ground for the citizens of Gates's digital nervous system. Teachers train students to be producers, users, and consumers for what is typically seen as the inevitable future. The very young are an especially precious commodity in this process, for the pathways that make up the adult nervous system can best be influenced in the young. But because young and old alike can participate in multiple ways in this (brave, new, largely commercial) world, we are all targeted for (re)training. As potential producers, we are trained to produce software, hardware, and applications (such as Web pages or product designs). As potential users and consumers, we are trained to purchase and use the latest software and hardware. In addition to being taught to be "computer literate," we are increasingly being taught to be "visually literate." Learning *how* to read the visual media landscape has become a priority academically, but even more significantly, learning *to* read it has become a priority. Training students to be "critical consumers" is first and foremost training *to* consume.

The shift in emphasis to new media (especially computer) development and use entails an almost necessary shift in, if not away from, the humanities and social sciences. There are, after all, only so many hours in a (school) day. On almost any given day, with a little ingenuity, you can find evidence of this shift in the balance of resource allocation. Recently, for example, National Public Radio's *All Things Considered* (Chicago Public Radio: March 23, 2001) reported that Loyola University in Illinois, a Jesuit institution, was eliminating its Classical Studies Department in which Greek, Latin, Ancient Literature, and Philosophy have been taught since 1870. In the report, James Keenan, Loyola Professor of Roman Law and Social History, expressed concern that Loyola was becoming "a kind of vocational enterprise." By shifting resources to fit the demands of the market and neglecting the cultivation of other capacities, we risk diminishing if not outright losing the very knowledge and skills that "keep us on our toes," that is, the critical ability to assess where we have been, where we are, and where we might go.

Clearly, people undertake these shifts in emphasis and resources believing that by retraining producers, users, and consumers for the digital universe they are preparing people for an inevitable and hopefully better future. They tend not to notice that what they are actually doing is *making* a future by *making* the people to fit its image. This error, and I do think it is an error, reflects a deeply entrenched habit in the West with respect to our posture on technology, and its effects are considerable. Having taught an (evolving) course in culture and technology for over 20 years, I can state with confidence that the commitment to equating both the development and use of new technologies and the demands of the market with the public interest is largely habit, by which I mean that the belief in their equation, and the conduct that reflects that belief, are enacted largely without reflection. The fact that these beliefs and actions might be merely (but powerfully) contingent on particular cultural arrangements, and might therefore be otherwise, has simply not occurred to most people. When alternatives are made visible, the "lights go on" in people's thought and conduct.

Thinking Geometrically "turns on the lights." It suggests that looking to new technologies, particularly the computer, to sort out the chaos of contemporary life is merely to contribute to the problem. After all, our Franklin Planners and Palm Pilots do not help us *think;* they merely help us *manage more* of the increasingly unmanageable. Rather than prepare for the future by shifting resources to support the industry of new technology development, literacy, utilization, and specialization, and producing (training) people to suit that industry, Waisanen would have us look in another direction. In *Thinking Geometrically* he asks that we pause and think through what it is that we *really* need. If the world gets out of hand because of the multidimensional, compartmentalized demands it makes on us, would not the skill we need to "hold it all together" be precisely the skill of multidimensional thinking? It seems so obvious. Why then do we corral people down the road of specialization, computer literacy, and new technology production and utilization, seeing this as an appropriate response instead of one that contributes to our inability to negotiate the complex contemporary landscape? Again, the error we make is to substitute the means for the ends. The new technologies (the means) are supposed to help us create a better future (the ends). However, we tend not to ask whether or not the future they are bringing us is better, or in what sense we mean "better." Rather, we tend to take on faith the position that the new technologies are themselves an indication of a better life. *Thinking Geometrically* whittles away at these habits of thought and conduct in contemporary culture by refusing to accept as a matter of faith a technological answer to what is a much larger question: What sort of thinking tools best serve the practice of consciously designing the future? In essence, between every line, *Thinking Geometrically* raises the question of wisdom.

This landscape of contemporary culture, then, constitutes the first part of Waisanen's definition of thinking geometrically: the "multiple and simultaneous dimensions of complex and variable objects and relations." It describes the world we have to work with, what Waisanen sometimes refers to as the "imaginative constructions and events in real space" that map our movement through everyday life. But thinking geometrically also involves a very particular way of "holding it together," as expressed in the second part of the definition: "in a structured, yet dynamic and interactive fashion . . . to construct a mental model, to compare this model to real events, and to modify both the model and one's own behavior to arrive at a better fit between imaginative constructions and events in real space." This form of "holding it together" differs in crucial ways from the scheduling and management "solutions" offered by the planners and PDAs (personal digital assistants) of contemporary culture. Geometrical thinking as the form of "holding it together" has in Waisanen's argument three parts (or dimensions): multimodal thinking and practice; multidimensional thinking and practice; and design practice of as a kind of lived wisdom.

To some degree *Thinking Geometrically* uses the terms "multidimensional" and "multimodal" thinking and practice interchangeably; they are inseparable dimensions that must be held together in the mind's eye to understand what Waisanen intends by thinking geometrically. Nonetheless, there is an advantage to looking at the different work performed by each of the terms, for each draws attention to a different view of the challenges before us. Multidimensional thinking is the more inclusive term, so I will work toward its elaboration by considering first the terrain covered by the term "multimodal."

Multimodal Thinking and Practice

In communication theory, the "modes of communication" argument asserts that the dominant mode of communication in any era shapes human culture in a central way. Typically, the dominant modes have been understood, in order of their development, to be orality, literacy, and electracy (or the electronic mode, information mode, or digital mode; it is always most difficult to name the mode in which you find yourself). The most influential voice in the argument for the changing modes of communication remains Walter Ong, who in his classic *The Presence of the Word* (1967) tracks what he calls the "shifting sensorium" as we move from one mode to another. He describes the shifts from a culture based on oral/aural, face-to-face interaction, to a literate culture based on the separation of messages from their senders and the use of these "denatured" words in exercising control over space, and finally to the "bewildering" and "elaborate indirection" of the electronic era (89). While arguments about the nature of what Ong called the "electronic era" have been dramatically updated to respond to the rapid development of the computerization and informatization of the world, these developments have only augmented the claim that the mode of communication dominates any particular age. The commonplace claim that we are in an "information age," characterized by a new organization of virtually everything, has its roots deep in the assumptions that a mode of communication characterizes an age, that the mode determines the shape of the age, and that a new mode eclipses and transforms old modes.

In visual theory, the modes of communication correspond roughly to the modes of spatial organization advanced by Henri Lefebvre in *The Production of Space* (1991) and include absolute space, abstract space, and differential space. From the Renaissance to the early 1900s, we lived, according to the argument for "modes," in literate, abstract space where "the word" dominates, where space is neatly compartmentalized by Cartesian coordinates, and where perspective is used to negotiate spatial relations. In the realm of abstract space, representation of "things" by "words" takes center stage. Living in the world becomes a "relationship" that is "mediated" by inscriptions, that is, by words, numbers, and symbols. In the transition from absolute space, "access" to the visual and to the "story space" of oral culture is restricted by a requisite mediation by the word, by representation. The visual and oral are thus diminished; they become subjugated to literacy, in both theory and practice.

What Waisanen hopes to find in the opportunities presented by the current transition to differential space, the "space of plurality and difference" (16), is the possibility of revisioning the world by encouraging access to the visual as unfettered by the limiting representations of inscription. That is, where once abstract space necessitated that we "read" pictures through criticism, we ought to be able equally to "see" criticism through pictures. In *Thinking Geometrically*, Waisanen emphasizes a restoration of the visual as primary and irreducible to the word, but his argument can easily be extended to a restoration of the oral/aural mode as well. To become fully multimodal suggests giving equal weight to thinking and living in visual, oral, and literate modes.

To think and live multimodally, then, implies a certain multidimensionality, for to think and live placing equal importance on the visual, the literate, and the oral/aural would be to hold all three dimensions in the mind's eye simultaneously, without subjecting one to the other.

Multidimensional Thinking and Practice

Multidimensionality, however, means far more than (just) multimodality. Multimodality provides entrée to a multidimensional universe, but it does not exhaust what you find when you get there. Multidimensionality resists codification because the "multi" implies infinite possibilities, infinite dimensions. One cannot circumscribe these dimensions with the categories sometimes supplied by theorists—such as communication theorists' "orality, literacy, and electracy"; cultural theorists' "gender, race, and class"; marxist theorists' "politics, economics, and ideology"; and so on. To each of these systems of classification, there is always something that can be and should be added. To the marxist formulation, we might want to begin by adding "technology and environment";

to the culturalist formulation, we might want to begin by adding "ethnicity and affect." And to each of these categorizations, we might want to add the others, such that "gender, race, class, ethnicity" and "affect" might be added to "orality, literacy, and electracy," which might be added to "politics, economics, ideology, technology, and environment." Each of these is a dimension with which we might negotiate reality in meaningful and useful ways. But why stop there? What about adding, say, the dimension of history? Or corporeality? And so on. Where do we stop? And what justifies our stopping anywhere? As Geoffrey Bowker and Susan Leigh Star explore in *Sorting Things Out* (1999), classifications such as these—maps, technologies themselves—can never "fully capture the terrain," although they do "help us to find one's way" in however limited a fashion. To navigate the world in an unlimited way requires a willingness to admit new dimensions that are not always readily apparent. To return to Waisanen's definition of geometrical thinking, this kind of thinking requires a constant process of constructing models, comparing them to real events, and a willingness to "modify both the model and one's own behavior to arrive at a better fit between imaginative constructions and events in real space."

The systems of classification within which we tend toward complacency—Bowker and Star's "dynamic compromise" (54)—is what Waisanen would call a "perspective." A perspective always selects and deselects, distorts, and compromises. In contemporary theory, in abstract space, in the literate mode, we tend to accept the "perspective" as the necessary tool with which to manage reality. When we are especially "enlightened," yet still tied to abstract space and the literate mode, we acknowledge the work of selection and distortion and argue for moving between one perspective and another to look at matters from numerous perspectives, that is, to look first from one perspective and then another. We might look, for example, at a simple classroom conflict between a student and a teacher as a problem of clashing perspectives. From the teacher's perspective, the problem might be one of a student's failure to commit to the work of learning. From the student's perspective, the problem might be one of having to set priorities with limited time and choosing to prioritize the classes that will most help on the job. As a problem of perspective, each perspective is equally plausible, yet each forecloses the legitimacy of the other. This paradoxical dilemma of perspective is described visually by Waisanen: Any particular visual perspective, of a building for example, so distorts the shapes, lines, and relations of the building that other possible perspectives become irreconcilable from that perspective. In other words, perspective always privileges one view at the expense of another; it is not equipped to do otherwise.

In contrast, Waisanen does not accept perspective as a necessary or sufficient practice or metaphor for either thought or everyday life. His reasons are built on the exposure of perspective as a nonnecessary, contingent, historically specific practice. Perspective has a history; it was not always the way we saw things; and it is currently under intense challenge in the transition to differential space. Moreover, perspective is something that we can learn to resist. Waisanen's goal is to advance in its stead a science of space in which multiple "views," rather than perspectives, can be constructed and "thought" in such a way that each view maintains the integrity of the others. Again his model for thought is visual, its possibility revealed in the work of descriptive geometry. Descriptive geometry, specifically orthographic drawing, permits "accurate" depiction. In practice, that means that each "view" is drawn in such a way that other views are truly represented in it. While a full picture is never possible (remember the dimensions are infinite), drawings can shift between views without one view distorting or disallowing any other. How, we are challenged, do we make that a model for thought and life?

Waisanen does not completely satisfy my desire for an answer to that question, but he does point to the kinds of skills we can cultivate to develop the ability to hold multiple views together in "the mind's eye." To *understand* contemporary space, but more important, to consciously design differential space, we need to develop equal facility with the multiple modes that operate—the oral, literate, and visual—and we need to be educated in ways to hold multiple views together. Because the visual has been the most neglected mode in contemporary culture, and because the study of the integration of multidimensionality is hardly a priority, his advice makes sense: learn to draw. I think there are two interrelated components to Waisanen's prescription to learn to draw: one requires hands-on practice working in the visual dimension, the other requires the study of the problem of dimensionality.

The first component of learning to draw, which emphasizes actual hands-on practice, demands active engagement with the visual dimension. Yes, we live in a very visual culture, but our role is primarily as recipients of previously constructed visual messages, not as producers of a potentially richer visual realm. We cannot learn to fully negotiate the visual dimension without direct bodily experience in it. Drawing demands bodily engagement by requiring us to think and work in multiple dimensions simultaneously. It involves not just the brain and literary practice, but the integration of a much fuller range of bodily capacity and knowledge. We can read a book about drawing, but it is not until we struggle with the practice of drawing that we learn to "see" in and through this way of engaging the world. Only then does the potential to design through drawing open the infinite potentials of a larger world.

Furthermore, as we learn to draw it is almost inevitable that we will find it necessary to handle the problem of dimensions. It happens in many ways: through reducing multidimensional life to the two dimensions of a page, through trying to achieve the sense of "surround" on a flat surface, through attempting to re-create the "experience" of an object that involves multiple views when only one view can be executed, through learning to create a drawing "in perspective," through experiencing the flatness that results from copying a photograph, and so on. While it is theoretically possible to learn to draw without facing the problem of dimensions, it seems unlikely. When Waisanen writes that he is not referring to the production of "fine art" in his prescription to draw, he might be referring to the tendency to copy technical moves to make a picture or the tendency of some artists to repeat what they have produced before with little or no further exploration of new dimensions. Exploration of dimensionality seems to be key here and it is difficult to practice drawing without having to explore bodily the challenges of dimensionality.

The second component of learning to draw—the study of the problem of dimensionality—directly addresses the need to explore the treatment of dimensionality as a contingent practice. In addition to urging us to "learn to draw," Waisanen suggests in several places that we "study graphics." With this statement he urges us to study the history and practice of drawing because to do so demystifies allegiance to "one right way" of seeing and living. I recall when I decided it was time to learn proper perspective so that I could render my paintings "correct." I was astounded when the book I read covered multiple systems of perspective, not one proper way, and that each of these systems involved unique forms of "distortion." That was an invaluable lesson for me as an artist; it freed me to explore and draw in ways that suit my own desire to *design*. But even more important, it helped me think about the limitations of the very notion of perspective as a normative concept with which to negotiate life. Studying graphical systems, like studying Waisanen's book, can open opportunities for each of us to see the need and the path to hold multiple views together.

Not all drawing practice and study is equal, however. Some forms stimulate the mind and body better than others. For Waisanen, the study and practice of orthographic drawing is a special kind of drawing, the kind most likely to stimulate the mind and body to develop the capacity to think geometrically. This is because orthographic drawing insists on maintaining the integrity of multiple views even as it privileges one over the other in reducing a drawing to two dimensions.

Drawing is not the only path to geometrical thinking, though it may be one of the best paths. Because the terrain in and on which we might practice geometrical thinking is infinite, there are myriad everyday paths for getting there. Geometrical thinking is not a rarified concept for philosophers to juggle in a theoretical circus act. It is a concept for the everyday. Waisanen looks at Plato's allegory of the cave as revealing something about geometrical thinking;

but so too does he look at Robert Ludlum's spy thriller, *The Bourne Supremacy* (1986), in the same light. Geometrical thinking is applied here to shipbuilding and carpentry as much as it is to orthographic drawing. Characters who can think geometrically are as likely to be the Jason Bournes of spy novels as they are to be engineering designers. Frankly, I suspect that whatever helps one see and respect the integrity of multiple dimensions is a practice that would be supported by Waisanen. My friend Marilyn Urion recently suggested that geometrical thinking practices might include making clothes without patterns (an old and respected practice), building without blueprints (as my craftsman husband still does), and building engines and vehicles from scratch (as was common practice in the early days of automobiles; they were called "jokers" where I live). It is interesting that all these examples are of "old" practices; they are not new, convenient practices, in which the path between the brain and the hand is guided by a particular kind of inscription, that is, by other people's patterns. Ironically, geometrical thinking might be stunted in the process of the contemporary computerization and informatization of the world. As Waisanen suggests, computer work may actually be enforcing a perspectival approach that renders us ill-equipped to think geometrically. Waisanen's experience with the limitations of the computer as a design tool have been recently underscored by Frank Wilson who, in *The Hand* (1999), insists that drawing on computers minimizes the involvement of the body such that there are significant and undesirable effects. Quoting Karen Chelini, the director of human resources at LucasArts Entertainment, he notes that in the world of artists who have worked extensively on computers "you'll see a stiffness or a flatness, a lack of richness and depth" (257). Drawing and designing on the computer transforms the visual every bit as much as it transforms writing. This unpopular suggestion—that drawing and designing on the computer may be producing undesirable results—is a crucial challenge we ought to consider. But we will lose our ability to evaluate the challenge in proportion to our inability to draw and our lack of knowledge of graphics. So, pick up your technical pens! We've a world to design!

Design Practice as a Kind of Lived Wisdom

Thinking Geometrically is for Waisanen as much about designing cars as it is about designing life. Holding multiple views in the mind's eye ought to provide us with the tools for designing a good life, for living wisely. And it is on this matter that I must part company somewhat with Waisanen's argument. Waisanen seems to suggest that geometrical thinking is an unqualified good. But I'm not so sure. It is indeed very interesting that the initial extended example Waisanen uses to illustrate geometrical thinking at work is the (fictional) story of Jason Bourne, at various moments a Cold War spy, killer, commando, and uncontrollable maniac. Waisanen, fully aware of the unconventional nature of his choice of examples, suggests that "the conventional conclusion one might draw from this is that Bourne's action-oriented, geometrical thinking is generally at odds with the status quo" (25). But he oversimplifies the implications of this opening with a clever retort: "The association of geometrical thinking with physical action and violence, as demonstrated by Ludlum's story, suggests that cultivating it might be a formula for cultural disaster. In a stable world, this might be the case; in a dynamic and unstable world, however, it might be the only solution" (25). Because I take as given the fact that human social life is always unstable, or at least always has been, Waisanen's comment elevates "physical action and violence" to somewhat privileged status. He does, albeit briefly, appeal to an ethical principle for measuring out doses of geometrical "physical action and violence": "One who has power, and uses it, has to have a sense of responsibility: to know when to stand still, when to tread carefully, and when to act forcefully" (25).

Waisanen misses a potent opportunity here to cast the same critical eye on geometrical thinking that he casts on perspectival thinking. In his enthusiasm for and excitement about what is truly an enormous leap in how we understand multimodal, multidimensional space, geometrical thinking is transformed into something either necessary and therefore inherently moral or at the very least morally neutral. But I'm not so sure. There are at least two challenges to make to this position. First, learning to think geometrically might involve asking too much of the human organism. Referring back to the quotation early in this preface by Thomas Merton, recall his conclusion that the "frenzy of the activist neutralizes his [or her] work . . . because it kills the root of inner wisdom which makes work fruitful." We need to look more closely at this practice of geometrical thinking and determine to what degree it might resemble the frenzy that Merton draws attention to and what the planners and PDAs promise (unsuccessfully) to manage. Does the prescription to think geometrically only make sense if we accept as legitimate the multitude of conflicting demands placed on us? And does accepting the need to operate geometrically have the capacity to destroy wisdom and creativity? I know from my own experience that I feel the most creative and am most proud of the products of my creative energies when I have been able to block out much of the world, when I can be deaf to the multiple demands on me that never seem otherwise to abate. In my artwork, that has necessarily entailed blocking out "the word," something rather risky for an academic, but absolutely necessary in my case. That is not to say, however, that when I paint I am not making geometrical connections; it is just that they are very carefully circumscribed. It is almost as though the intense and focused depth required in the creative act cannot be sustained by the noisy breadth of too much geometrical connection. Similarly, when my husband, Kenny Svenson, designed and built our (extraordinary) timber frame house with the image held only in his mind's eye, he spoke very little. The depth of his concentration took him virtually "out of the world" for months. He simply could not have built our house without resisting making some geometrical connections in lieu of others.

In contrast, the sense of exhilaration on the job—managing telephone calls, computer messages, committee meetings, classes, students, writing, and lunch; mailing this document there, delivering this resource there, and so on—looks suspiciously like thinking geometrically but does not resemble wisdom; nor does it ensure a better life—for me or anyone in my path. Gates's digital nervous system also looks suspiciously like thinking geometrically; he is perhaps a Jason Bourne of a digital cold war. Gates's digital nervous system, and even my own lesser management skills on the job, do involve making creative geometrical connections. Clearly they do. But it would be a sad world indeed if Gates's model for business was to serve as the model of geometrical thinking and practice—and morality!—in the new world of differential space.

These contrasting examples suggest that at its most extreme geometrical thinking could, in the guise the contemporary culture is most likely to embrace (the Gates model), promote imbalance, an eye blind to questions of the relationship between wisdom and creativity. If that is the case, geometrical thinking might not be an absolute to strive for, as I believe Waisanen assumes it is, but a skill that we can employ at variable "speeds," in variable circumstances, with variable affects. It is perhaps better understood as a skill we have always had in more or less supply and which we apply differentially as the world changes. Moving into differential space may demand that we be better at it, but that reality comes with no guarantees of being a more desirable space to inhabit.

The second challenge to assuming the inherent morality or moral neutrality of geometrical thinking, related to the first, involves the proposition that knowing more about how the world works does not necessarily make one wise, automatically, as it were. In suggesting that those with the power to think geometrically need to use it "responsibly," Waisanen acknowledges the disjuncture. However, he makes very little out of what I take to be a matter of enormous import. I cannot imagine a more powerfully frightening despot than the one who can think geometrically. (Did not Hitler think geometrically? At least to some extent?) And where in the training to think geometrically do we train people to be ethical? In the "Philosophy of Technology" class I am currently teaching, one of the very best "thinkers" in the class, one of those who does indeed seem to grasp and hold in his mind's eye the work of multiple views, still plans to design war ships. I must admit to not understanding why. And I'm not sure that skill at geometrical thinking is enough to overcome the violence in war and in everyday life. I fear, in fact, that it is quite capable of accelerating it, just as Waisanen suggests.

So what else is needed? As Waisanen was working on this book I encouraged him to read the work of Gilles Deleuze and Félix Guattari. I suspected that their concepts of territorialization and deterritorialization would have provided him with a particularly generative way to conceptualize the constricting apparatus of, and escape from, forms of classification such as perspective. Although he intended to read their work, he decided to postpone the study until completing this stage of what was clearly (and prophetically) to be a life-long project. I suspect that Waisanen would have found delight in the work of Deleuze and Guattari, in which the "something else" that is needed to parse constructive from destructive geometrical thinking is provided, at least for me. They too resist the confines of limited dimensional (like perspectival) thinking, although it is understood by them as a process and product of *territorialization*. In contrast, they privilege (like Waisanen's multidimensionality) deterritorialization (Deleuze & Guattari, 1987). But they are explicit about the fact that deterritorialization can lead to danger, for without structure, classification, and/or some semblance of a (necessarily limited and limiting) "perspective," we might either descend into madness (like Jason Bourne) or become uncontrollably violent (like Delta, another of Jason's "dimensions"). Deleuze and Guattari consider the possibilities in terms of the four dangers: "first, Fear, then Clarity, then Power, and finally the great Disgust, the longing to kill and to die, the Passion for abolition" (227). Geometrical thinking, it seems, opens us up to all these risks. Fear is the fear of losing grounding in known territories that leads us to cling to "security." We can fear the demands of geometrical thinking and cling to the old notions of abstract space—the familiarity, manageability, and comfort of "perspective" for example. Clarity is the appearance of and reveling in the new suppleness provided by processes of deterritorialization. Deliriously, we can celebrate the delicious clarity of infinite dimensions and possibilities and go no further than indulging their proliferation. Power is the ability to shift and "stop" the proliferation of possibilities to suit particular ends. We can use our knowledge of multiple dimensions in the exercise of power, like Gates's use of geometrical thinking to gain wealth, advantage, and power in the marketplace. Finally, the great disgust is almost a perverted form of celebration that arises from a cynical embrace of infinite dimensions and possibilities. This disgust leads to a perpetual revolutionary overturning of any consistent place to stand—a line of death, a passion only for destruction. It is possible to let the demands of geometrical thinking "require" that we overturn everything we once knew to be "true" with a zeal that can only result in violence and destruction, where any place to stand is denied in zealous despair.

There is a dark side to geometrical thinking, of that I am sure. So, be forewarned and cautious as you work your way through this delightful and generative book. This book will provoke you, but it will not and should not answer all your questions. It does not offer a final, finished map of the territory. I suspect that thinking geometrically, looked at from a Deleuze-Guattarian perspective, could not produce such a map. Rather, the struggle is the destination.

A Note on John Waisanen

I've edited this book because John Waisanen died of non-Hodgkin's lymphoma on March 29, 1998, before he was able to undertake the process of revising his manuscript for publication. A bit of the extraordinary story needs to be told for several reasons. First, the circumstances in which Waisanen persevered, and completed, the manuscript exemplifies his commitment to the importance of thinking geometrically. Second, as a reader you might understand the book better if you share in some of the dilemmas I faced as its editor. And finally, this note provides the ideal opportunity to honor my friend and colleague by sharing the personal side of the project in which its author must be "John," not the academic impersonal, "Waisanen."

John's unique career path kept him moving back and forth between industry and the academy. As a successful trainer of engineers and designers on the use of computer-aided design (CAD), he encountered questions that kept driving him back to research in the academy. When I met him as a graduate student in Michigan Technological University's Department of Humanities, he was searching for a way to understand the significance of what he experienced as the widespread inadequacy of people's ability to maneuver in the visual realm. His concerns were rooted in the practical everydayness of the problem, and he was often frustrated by academic "readings" of the visual terrain that missed the crucial work of the visual in negotiating movement through everyday life and work. After all, lack of visual skill in the design process can result in automobiles that won't run, bridges that collapse, buildings that crumble. His work was thus motivated by a deep ethical drive, and he researched his way through a variety of literatures to find his way to a solution. Shortly after he began to piece together his argument, he became ill, but he kept writing. And as his illness advanced, he kept writing. His commitment to the work is best illustrated by the fact that he finished writing the final chapter on a laptop in his hospital room while undertaking massive radiation and chemotherapy in preparation for a bone marrow transplant. I kept reminding him that he didn't need to do this, that if he wanted to stop writing—even abandon this project—it would be okay. I asked Susan, his wife, if she wanted me to insist that he stop, but she assured me that the work was a really important part of his life and that it was good for him to continue. After his dissertation committee read the completed manuscript and offered suggestions for revisions, John was able (astonishingly, in one of his last stages of coherence in his battle with infections for five months after the bone marrow transplant) to complete one full revision before he died. He had always intended the manuscript to become a book, and he and I talked at length about what form those further revisions might take.

My familiarity with John's commitment to his argument in *Thinking Geometrically*, coupled with my knowledge of his intended revisions, left me the ideal editor of this work. But, somewhat problematically, as I had become deeply convinced of the significance of John's argument, I became equally passionate about reaching out with this argument to make a difference in the world. My investment in the argument has rendered me something less than the ideal editor. I have wanted to continue my conversations with John through this book, changing things as I thought he might as a result of those conversations. As a consequence, it became very difficult to distinguish between necessary editing and intrusive revision. I have tried to remain an editor, not a coauthor.

As a consequence of my dilemma as part editor, part coconspirator, I have chosen to leave intact some elements in the text that I am confident John would have attended to down the road. I have addressed some of these matters, above. Specifically, I suspect that John would have clarified the distinction and relationship between multimodal and multidimensional. It would also have been easy to convince him to address the ethical matters more directly. Most pointedly, there would be more examples of thinking geometrically and a more explicit statement of what a science of the visual might look like—if John had had more time. I could not do those things for him. The lack of examples and the absence of at least a "bare bones" statement of the shape of a science of the visual is a sad lack. But in defense of *Thinking Geometrically*, John lays the groundwork for us to do that work ourselves. And doing the work ourselves is very much what John was after. He was, after all, not only a scholar and CAD trainer; he was an antique collector and restorer, an instrument builder and restorer, a superb banjo and fiddle player, an artist who used his technical pen in balance with his work at the computer, an integral part of his church and community, and a deeply loved father and husband. He lived geometrically. And for him living geometrically is, in the end, what matters.

CHAPTER 1

PURSUING MULTIMODAL SPACE

This project actually began more than 20 years ago, though I didn't know it at the time. As an undergraduate at Michigan State University, I struggled to find a theoretical home from which to address issues of visual communication that seemed to fall between the cracks of existing disciplinary and intradisciplinary boundaries. I eventually settled on communication as a major. Being exposed to the theory and applications of multidimensional scaling served as a catalyst for my interests in the visual and spatial representation of information, but I was never as interested in the quantitative approaches to communication as I was in the visual aspects of it.

After graduation, and a brief and undistinguished stint as a graduate student in communication, I married, relocated to metropolitan Detroit and fell into a career in automotive design, technical illustration, and computer graphics. A heretofore unrealized and untapped talent for visualizing and solving spatial and graphic problems enabled me to advance. With training in descriptive geometry, and experience in both manual and computer-aided design (CAD) at the Ford Motor Company, I moved to Prime Computer, Inc. where I was responsible for developing courses and training engineers and designers in the use of CAD. Simultaneously, I resumed academic studies and completed a Master of Arts degree in industrial design at Wayne State University. Rather than backtrack (financially, at least) into a completely new field, I instead advanced into project management for technical documentation, specifically, the design and production of CAD user manuals, a position I would never have expected to be in several years earlier.

Throughout the nearly two decades of this varied academic and practical experience, I was continually puzzled by what seemed originally to be a simple question: Why was the ability to accurately imagine, visualize, or represent objects in space or spatial relations so uncommon, even among trained professionals in fields like design or engineering, disciplines that required this ability as a condition of success? A surprisingly large number of engineers with whom I worked, both as a designer and trainer, really couldn't "read" the drawings they signed; nor could many designers supply a decent three-dimensional (3-D) illustration or sketch of their work.

The simple answer to this question is that they simply lacked the talent or ability to envision either objects or spatial relations. This ability, however, can be cultivated through hands-on experience and training with machinery and drawing techniques that used to be, but are no longer, standard background for designers and engineers. In producing CAD documentation, developing courses, and delivering training, I struggled to find better ways to demonstrate and explain to others what was intuitively obvious to me. I found that words alone were rarely enough, and pictures didn't help those who couldn't make sense of them in the first place. I began to sense that the real problem (or set of problems) was deeply rooted in cultural priorities and relations that were beyond my own immediate grasp.

The industrial priority of getting things done on time and within budget left little time for systematic inquiry into these personal and philosophical concerns. Perhaps knowing of these interests, a technical writer I had hired mentioned that Michigan Technological University (his own alma mater) had recently begun a doctoral program in rhetoric and technical communication. I was ready to make a career change; I applied for admission and received a three-year fellowship. On the basis of this theoretical affirmation of the importance of my interests in visual communication (and the guarantee of but a fraction of my former income), I left industry and went back to school with a new sense of purpose.

Soon after I began my studies, the real complexity of my concerns became apparent. The "simple" question of why most people have trouble with visualizing graphic information, and my interest in developing "better" ways of communicating such information, were far more problematic than I had imagined. I found that the relation of technical communication, both to other academic disciplines and to culture in general, was itself the focus of intense debate in which words, not pictures, were the primary tools. In this context, it was not only difficult to define my own theoretical home; it was also difficult to explain to anyone what I wanted to do. I was convinced that "thinking geometrically"—a notion drawn from Robert Ludlum's novel *The Bourne Supremacy* (1986)—held tremendous, untapped potential for understanding not only the problem of visualization, but how we act in, and are acted on by, the world. Further, I was convinced that orthographic drawing, as demonstrated in engineering graphics, offered a formal path in the development of geometrical thinking.

I persisted, read widely, and gradually realized that I was dealing with what are, fundamentally, problems of space. Henri Lefebvre's monumental work

The Production of Space (1991) provided me with a broad theoretical and historical framework for understanding how social relations, technologies of representation, and structures of knowledge are not only based in, but *produce*, cultural (not just mathematical) conceptions of space. Changes in cultural "modes of space" throughout history correspond roughly to the "modes of communication" described by Walter Ong in *The Presence of the Word* (1967); however, despite the centrality of "the word" in human affairs, language serves poorly in the investigation of problems of space.

The priority of language in the construction of knowledge tends to overshadow the power of pictures, which resist explanation in words. Discovering the work of W. J. T. Mitchell, *Picture Theory* (1994) in particular, enabled me to better understand the complex and interpenetrating relationship between words and pictures in the production of meaning and space. Duncan Davies in *The Telling Image* (1990) called my attention to the dominance of pictorial technologies in contemporary mass media, as well as to the widespread lack of the necessary cognitive and practical skills to sort and organize an increasingly graphic stream of information. Bruno Latour's essay "Drawing Things Together" (1990) made me aware of how the imagination, demonstrated by the example of Galileo, works through graphics, words, and numbers. As "inscriptions"—knowledge made visible—graphics, words, and pictures both produce and embody scientific knowledge. This, in combination with the work of William M. Ivins in *Art and Geometry* (1946) and *Prints and Visual Communication* (1953) and Hubert Damisch in *The Origin of Perspective* (1994), led me to recognize the central importance, and limitations, of perspective as a model for thought in producing the space of contemporary culture. Edward Soja in *Postmodern Geographies* (1989) and José Ortega y Gasset, who also uses Galileo as an example in *Man and Crisis* (1958), provided models for understanding the joint character of the historical and spatial imagination. Both, I came to see, are exercises in geometrical thinking.

Edwin A. Abbott's story, *Flatland: A Romance in Many Dimensions* (1952), clarified the geometrical concept of "dimensionality" in human knowledge and communication. This story, together with passages from Plato's *Republic* (1957), especially the allegory of the Cave, led me to understand how the human imagination is both inspired and constrained by cultural conceptions and representations of space.

Peter J. Booker's *A History of Engineering Drawing* (1979) provided an overview of the role of graphic technologies throughout the course of Western history. The advancement of these technologies has been for centuries (and still is) hindered by a deep division between the empirical and the theoretical development, and application, of graphic theory. Despite the slow diffusion of graphic knowledge across disciplinary boundaries beginning in the early

Renaissance, arriving at the understanding of drawing as projection opened up new ways of representing and thinking about space and problems of space. Geometric perspective came first, but the culmination of drawing-as-projection occurred with Gaspard Monge's invention of descriptive geometry in the late eighteenth century. Descriptive geometry provides a theoretical basis for orthographic drawing practice; even after 200 years, it is still poorly understood despite its central importance in modern science and technology.

Recent work in the emerging field of design studies by such theorists as Richard Buchanan and Victor Margolin in *Discovering Design* (1995) helped to expand my previous belief in the crucial yet largely unrecognized role of design in transforming both social and physical space and our conception(s) of space. Design ideas are communicated most obviously through graphics, but these are inextricably woven with verbal and rhetorical dimensions of thought and experience. Design is a potentially architectonic discipline, which may serve as the basis for the science of space called for by Lefebvre.

Weaving a meaningful pattern with all these rather different threads was difficult; I expect that the result will similarly challenge the reader. I struggled for years with it all until finding the key that I needed to assemble these disparate and resistant threads into a coherent structure. I found this key (ironically) in *The Literary Mind* (1996), in which Mark Turner defines the parable, the process of condensing and projecting a large and complex story into a small space, as fundamental to human thought and knowledge. Through parable the human imagination constructs projective—spatial—relations not only between parables and subject matter, but across parables themselves. These relations are often ambiguous, elusive, and difficult to study through language alone. Mathematics is well suited to defining or solving problems of physical space, but it serves poorly for problems of social space, such as those history tries to account for. I realized that orthographic drawing, like perspective, makes projection more explicit than either language or mathematics and may well serve better than perspective as a model for thought in the pluralistic, multidimensional, and dynamic space of contemporary culture.

Words, numbers, and pictures constitute different planes of knowledge and experience, but to assert the dominance of any one of them in explaining problems of space serves only to limit one's comprehension of the complex space of greater dimensionality that they jointly inscribe. Words, numbers, and pictures are collaborators, rather than competitors, in producing meaning. In using them to exercise the human imagination—to envision—we all engage in the design of space. Insofar as we generally lack the disciplined ability to use words, numbers, and pictures with equal effectiveness throughout everyday life, the spatial problems, inequalities, and inconsistencies of contemporary culture will remain only partially understood and poorly accounted for. This is a historical problem; as long as we remain unable to account for such problems, we remain condemned to live with and repeat them.

I know now that the "simple" question I started with—why it is that so few people can visualize—has no simple or final answer. However, part of the problem is clearly the long-established dominance of spoken and written language as the basis for the construction of knowledge. Although language and mathematics in a sense compete with each other for priority in the construction of knowledge, much as the humanities compete academically with science and technology, they work together to inhibit the investigation of the potential that graphics offers as a path to knowledge. As long as this is the case, pictures will remain problematic, and an increasingly graphic information environment will become increasingly unmanageable. Language and mathematics are not the most appropriate tools for dealing with problems of form and spatial relations. Seeing—representing and solving problems of space—is in many cases best done through graphic means. Writing is the language of history, just as mathematics is the language of science. Graphics has yet to come into its own, but graphics serves as a better basis than either words or numbers for developing a language of space.

Problems and contradictions of space cannot go unrecognized forever. They may seem to fade into the background, but in reality they just take a different form. Having everyone learn to draw, or learn geometry, won't really correct the fundamental problem. Just as words and numbers are necessary but insufficient by themselves, so too are graphics. The goal I pursue in this book is to learn to recognize that words, numbers, and pictures are equal collaborators in the construction of a multimodal space. We can accomplish this recognition through thinking geometrically. Cultivating and exercising the ability to think geometrically on a widespread basis is a necessary step in designing space. To explain what I mean by thinking geometrically is one of the main tasks of this book.

In this introductory chapter I have intentionally avoided summarizing the structure of this book. This is only in part because of the sheer difficulty of doing so. More significantly, however, the questions and issues raised here are more complex than I had imagined. Indeed, they seem to "project" in multiple directions. What I have tried to do instead is sketch the motivations, questions and resources from which this book emerged. My goal has been to develop what Lefebvre would call a "space of representation," which permits, even encourages, multiple interpretations. The story of that space, as I construct it, is but one of many possibilities; each reader will no doubt draw from it other possibilities. This individual process of discovery is essential to understand space and to develop the geometrical imagination. I hope that this multifaceted story will challenge the reader to do so.

All representations of space are dimensionally limited; they are fixed in space and time. Relying on either textual language, mathematics, or graphics as the primary language to understand or solve problems of space is ultimately a self-limiting exercise. Why? Because space must be (is) *lived.* Geometrical thinking involves the recognition that each and every one of our actions—from a random act of kindness or writing an essay to executing grand schemes of world domination—is directed at some level to either reproduce or to modify the character and quality of the space in which one lives. What works once might not work twice, because the context is no longer the same. Space is multidimensional, dynamic, and unstable. Through thinking geometrically it may be possible to arrive at a multimodal language of space to serve as the basis for a humanistic science of space such as called for by Lefebvre.

Simply becoming aware of the power of thinking geometrically is but a first step toward making the most of one's involvement in, and responsibility for, the social production of space. Taking advantage of the conceptual tools that descriptive geometry and orthographic drawing offer toward developing the ability to think geometrically is another step. The practice of geometrical thinking involves design, not only in planning and choosing a course of action, but in discovering where and how one can get the most leverage given the resources at hand. Recognizing that any one of us can change the world by cultivating the ability to think geometrically presents a new set of social and spatial problems; thinking geometrically is no guarantee of rightness or goodness, but in the long run, I believe that geometrical thinking will lead us to pull together rather than pull apart.

CHAPTER 2

THINKING GEOMETRICALLY

Don't know much about history . . .
Don't know much trigonometry
Don't know much about algebra
Don't know what a slide rule's for
But I do know one and one is two
And if I could be the one for you
What a wonderful world it would be.
—"Wonderful World"
© 1960 by Sam Cooke

Problems of Space

Geometry was not immortalized in Sam Cooke's list of "don't knows." It certainly ought to have qualified, just as history, trigonometry, and algebra did. To anyone who ever took a geometry course, thinking about the subject probably evokes distant and even unpleasant memories: long moments of staring at a smudgy blackboard jungle of Greek letters, equations, and wireframe sketches of intersecting arcs and polygons, all somehow tied together in some theoretical space . . . and *just not getting it.* To me and, I would guess, most of my high school classmates, what little we grasped of geometry was its abstract character. It not only seemed disconnected from other subjects, but it was completely foreign to the concrete practices, priorities, and trajectories of our everyday lives. As adolescents in the Age of Aquarius, we were interested more in exploring the new morality than we were in numerality. After all, wasn't love the answer? Though, I don't think we really got that, either.

Not only did we fail to grasp individually the primary subject matter of geometry, we also failed to grasp collectively any sense of the historical or contemporary importance of geometry. Applied geometry was probably first developed by the Egyptians to reconstruct fields after annual destruction by the nourishing Nile floods. What we now call classical geometry was a Greek invention, usually attributed to Thales. Greek geometry reached a theoretical peak in the work of Euclid, but the practical origins of geometry remain forever embodied etymologically in the Greek roots of its name: *ge*, earth; and *metria*, measurement.

Euclidean theory has served both to inspire and constrain Western geometrical thought for over 2000 years. Even though the limitations and inconsistencies of Euclid's work were finally understood when modern theorists conceived non-Euclidean systems of geometry not much more than a hundred years ago, his work is still the basis for most introductory courses in plane and solid geometry.

Modern geometries are even more theoretical in character and more removed from everyday life than classical Greek geometry. Most of us, including me, were never formally exposed to these systems, since they are taught mainly in upper-level college courses impenetrable by any but math, science, or engineering majors. As a result of this lack of exposure, contemporary cultural perceptions of geometry and of problems of space are still rooted in the fixed identities and stable spatial relationships or structures that characterize Euclidean thought. This perception is outdated. To mathematicians today, geometry is no longer the science of measuring objects and relationships in space, as it was in classical geometry; it is now considered to be the science of transformations (Kramer, 1970).

In a multidimensional and dynamic cultural milieu where transformation is perhaps the only constant, modern geometry may be more appropriate than Euclidean geometry as a model for thought. However, perhaps because modern geometry relies on largely mathematical rather than graphic methods,[1] it is even more difficult to understand and apply than Euclidean geometry. Developing the untapped potential of geometry as both a theoretical model and as a tool for practical thought and action demands a resurrected awareness of the fundamental role of the visual and graphic imagination in geometrical thinking.

The serial or linear thought processes developed through literacy and numeracy serve the geometrical imagination poorly. Mastering the subject matter of either Euclidean or modern geometry requires developing a proficiency in what cognitive scientists call "parallel" thinking, which is largely based in imagery (Kosslyn & Koenig, 1992). Language doesn't offer a very good vocabulary for describing such thought; nor does such thought reduce easily to mathematical expression.

"Thinking geometrically" is the ability to deal imaginatively with the multiple, that is, numerous and simultaneous, dimensions of complex and variable objects and relations in a structured, yet dynamic and interactive fashion. In a sense, this means being able to construct mental models, to compare these models to real events, and to modify both the models and one's own behavior to arrive at a better fit between imaginative constructions and events in real space. People do this kind of visualizing intuitively all the time at many different levels. But without a suitable and visible language for systematizing such thought and for communicating these thoughts to others, thinking geometrically must remain a mysterious and problematic activity.

There is no one best way of thinking geometrically. Geometry itself is a diverse subject, a multidimensional field that permits multiple frames of reference, multiple points of view and multiple systems of representation. This multiplicity is not always amenable to mathematical expression, any more than it is amenable to description in words or depiction in images. In everyday life, often as not, spatial intuition—geometrical feeling—more so than formal rules, tells us how to choose one course of action, point of view, frame of reference, or system of representation over another. However, leaving decisions up to intuition leaves a lot to chance. The challenge of thinking geometrically involves learning to better manage one's intuition and imagination, and to better communicate the results of this mental effort in terms that are intelligible to others. I believe that the study of the visual, more so than the study of words and numbers, has much to teach us about the workings of the geometrical imagination. Pictures and the graphic practices involved in making pictures have to be understood on their own terms, through direct experience and knowledge of graphic communication, rather than being translated into verbal or numerical expression. As yet we lack the tools to do this. For example, Davies (1990:16) observes that even though pictures have today become "the main means" of the ordinary person's "reading," they are not yet a widespread means of "writing." Davies' metaphor—based in a notion that there might exist some kind of "visual literacy"—is a flawed and self-limiting position. As Mitchell (1994) points out, pictures may not be explicable through a model based on the metaphor of literacy.

However, as long as one resists the temptation to explain pictures through words or numbers, there is something to be gained through the metaphor of literacy. If one is to understand and take advantage of the power of pictures in the construction of knowledge, one must be able to both "read" and "write" them. Reading and 'riting are different sides of the same coin. Without proficiency in both, the so-called higher forms of thought made possible through literacy, like logic and philosophy, would be impossible, just as science would be impossible if all we could do is read numbers and not write them. Without some kind of appropriate formal language, such as the phonetic alphabet and Arabic numerals that make verbal and mathematical thought and communication available, graphic thought and communication—pictures—would continue to be problematic.

This is not to say that words and numbers ought to somehow be deemphasized to elevate pictures to a privileged position in the construction of knowledge. Rather, the study and application of graphic practices, guided by geometry theory, might provide the basis for developing a "language" of images. Having language is a requirement for developing other forms of thought—for example, either a philosophy or a science of graphics—that are impossible through words or numbers alone.

The representations we routinely construct through reading, 'riting or 'rithmetic lead to but a partial and limited understanding of the joint role of words, numbers, *and* pictures in the construction of knowledge. The historical emphasis in Western technological culture on literacy and literate forms of thought (demonstrated by such concerns as "why Johnny can't read") serves both to establish and reinforce artificial and unnecessary boundaries between these dimensions of knowledge and experience. Even numeracy suffers from the priority of language-based studies; it is a rare scholar who can read and write equally well in the spheres of words and numbers.

To most adults, being "numerate" means being able to do the number-juggling necessary to keep the bills paid and the checkbook in order. But just as being literate means more than the ability to read road signs or restaurant menus, there is a lot more to being numerate. Indeed, Western technological culture as we know it would not be possible without arithmetic, algebra, trigonometry, and calculus, but most of us are only minimally proficient in the first two and barely acquainted with the rest. On an everyday basis, we nonetheless utilize higher math without even knowing it. Advances made in the last hundred years or so—topology, group theory, and fractal geometry, for example—may seem far removed from the needs and practices of everyday life, but they are integral to the increasingly graphic technologies that are transforming our worlds of work and leisure. As long as we are unaware of these workings, we can scarcely hope to be aware of, much less become involved in or gain some measure of control over, the myriad processes (both mathematical and nonmathematical) through which our lives are organized and reorganized everyday. Such ignorance is an unaffordable luxury. But before we can begin to address this deficiency, the educational priorities of contemporary Western technological culture must be reexamined and reevaluated.

The traditional basics of modern education are supposed to be "the three R's": reading, 'riting, and 'rithmetic. The reduction to "R's" through the contracted pronunciation and spelling of the most polysyllabic and therefore difficult word—arithmetic—and the phonetic misspelling of writing is not simply a ruefully humorous testimony to our lack of systematic success in these dimensions of education. At most, the *three* R's must be said to constitute a *two*-dimensional education, one that prioritizes language training and rudimentary mathematics above all else. Reading and 'riting are hardly separate or isolatable activities. Together, they can be said to be related activities in/for a single dimension: literacy. Literacy is more than the ability to read. It also involves the ability to write and requires one to think in certain patterns that are unavailable to a purely oral mentality. Focusing on reading alone actually deemphasizes the importance of other cognitive skills that are components of literacy.

Ordinary students are usually well acquainted with various forms of literature—poetry, oratory, essays, novels, and encyclopedias—by the end of high school, even if they are not proficient in either using or producing examples. In comparison, arithmetic is only one small part of the field of mathematics. Typically, only honors students progress through algebra and trigonometry to geometry and calculus. If students are never required to advance academically beyond the rudiments of numeracy, we should not be surprised that literacy has such a privileged status. Literacy may be the only thing many graduates are left with after they've forgotten everything else they learned in school. Unfortunately, without other advanced tools and thinking skills, such as those that numeracy supplies, the minimally literate individual is barely qualified to enter the contemporary world of work.

Focusing on the dimensions of literacy (first) and numeracy (second) at the expense of other dimensions of education and experience—such as art and music, programs which are often the first to be cut in a budget crunch—is evidence of a widespread if fundamentally limited and shortsighted understanding of the kinds of thinking skills we need to develop through education. In an era when pictures are increasingly important as the primary means of mass communication (Davies, 1990), I would think that the formal study of visual (that is, graphic) communication ought to assume a high priority in curricula at all levels. The only subjects even close to the study of graphic communication that I can remember from school are art and drafting. However, after years of experience as an illustrator and designer, occupations for which the scant academic exposure I had served as poor preparation, I am led to believe that perhaps *representation* ought to become the fourth "R."

Unfortunately, "the *four* R's" just doesn't roll off the tongue very well. None of the first three are dispensable, however, and we don't have a single good word with which to condense reading and 'riting into one to get back to three. Nor do we have a word to describe the graphic equivalent of literacy or numeracy. The value of a good acronym or catch phrase in promoting a concept to the public makes the pursuit of one a worthwhile effort: Instead of the three R's, I offer the three D's: description, depiction, and differentiation. Even though this resonates well with the three dimensions of real experience, I'm afraid this coinage loses power, and wit, in its lack of brevity when compared with the three R's.

All irreverence aside, my point is that the widespread cultural and educational emphasis on literacy and numeracy leaves out too much of what we need in order to manage our lives in an increasingly technological and graphic environment. The widespread absence of formal experience with mathematics beyond mere basics (that is, training in arithmetic, and sometimes algebra or trigonometry) and the near-total absence of visual communication knowledge and experience among supposedly educated people is an omission that we can scarcely afford in either theoretical pursuits or practical affairs. Employers today are increasingly desperate for people who can manage effectively in an ever more complex and dynamic information environment. Neither scientific nor humanistic studies are well served by scholars who are unable to establish meaningful connections across narrowly specialized disciplines, or to bridge the gap between theory and practice even within one's own specialty. It is not only the quality of our lives (and livelihoods) that is at stake here; it is also the fundamental character and value of knowledge. As more and more people get out of school and just "don't know much about" history, or any other subject deemed important enough to be taught in schools in the first place, we are all the poorer for it.

Even though our daily lives may be centered on the key skills of basic literacy and numeracy, I believe that any curriculum designed to emphasize these two areas while deemphasizing other dimensions of knowledge will ultimately prove counterproductive. In particular, coping (theoretically and practically) with the increasingly graphic character of electronic communication technologies will be difficult. From the limited point of view of literacy and numeracy we may never be able to explain what Mitchell (1994) calls the "pictorial turn" of contemporary culture and information technology or understand the problematic relation of words and numbers to pictures. Nor will we be able to make any kind of prediction as to where current trends might lead even a few years from now, much less manage the cultural and technological changes that we have come to accept as an inevitable result of "progress." The only certainty is that more change will occur, and for anyone unable to keep up with the pace of change, the future is going to be anything but a "wonderful world."

These multidimensional social, economic, and political challenges might best be understood as problems of *space*. The study of problems of space is an exercise in thinking geometrically. That is why I begin with an investigation of the history of the joint role of words, numbers, and pictures in producing not only our knowledge of space, but also the fundamental character of space itself.

Producing Space

One of the key transformations involved in thinking geometrically is that of translating theory into action. The converse of this transformation is translating action into theory. The systematic means we have for accomplishing either of these translations are words, numbers, and pictures, which are conventional, static, and fragmentary representations—partial models—of physical or social space. A single, complete, or universally intelligible representation of the entirety of social or physical space may be impossible. Yet, some understanding of that entirety and how it is produced might be envisioned through the development of a geometrical model that considers words, numbers, and pictures as transformations, both of each other and of their subject matter, in multiple dimensions.

The multimodality of experience in the contemporary world invites multimodal, multidimensional thinking. Thinking geometrically is the ability to recognize multidimensional objects and relations in their dynamic multiplicity. They are always more than any of their possible two-dimensional and spatially fixed representations or identities, whether they are based in physical experience and accumulated knowledge, and/or expressed through words, numbers, and pictures. But, more important, thinking geometrically is also the ability to envision the possibility of alternate representations and to direct one's actions, and even those of others, toward the transformation of these actions, objects, and relations.

Thinking geometrically is more than just a way of better understanding technologies of communication that are increasingly visual and graphic in contemporary information media. The study of geometrical thinking may also lead to a better understanding of the joint role of words, numbers, and pictures—what Bruno Latour (1990) calls "inscriptions"—in the construction of meaning. According to Latour, inscriptions always embody more than their obvious and superficial content; they are also (rhetorical) instruments of knowledge and cultural power. However, neither can the workings of the human imagination be fully accounted for by focusing only on their rhetorical dimensions. There are other non-rhetorical motivations and forces at work in the making of knowledge.

Thinking geometrically, as I intend it, is a way of bridging the gap between the theoretical and practical dimensions of individual human imaginative activity in the production of knowledge. To get there, it is particularly instructive to investigate what Henri Lefebvre (1991) calls "practices of space" and their related "modes of space." The most convenient documentary evidence for this investigation consists of inscriptions: words, numbers, and pictures.

Lefebvre distinguishes between different kinds of space: for example, the physical, or natural, space of objects in the world; and the ideal space of mathematics and philosophy. He asserts that space is not something "out there," independent of human existence. Rather, space is a complex concept, produced through our action and involvement in the world. He identifies three distinct yet interpenetrating aspects of what he calls the "(social) production of (social) space": practices of space, representations of space, and spaces of representation. Inscriptions are but one instance of each of these moments. For example, writing and drawing are spatial practices that result in representations of some kind. Any concrete representation is a record, not only of its subject matter but also of the practices involved in its making and the frame of reference that served to guide its making.

According to Lefebvre, different cultures throughout history have produced distinct modes of space, characterized by different configurations of the three aspects of social space. Moments of significant cultural transition occur as modes of space change. Lefebvre identifies three such moments as being pivotal in the spatial history of Western European culture. The first is the emergence of Greek philosophy after the invention of the phonetic alphabet, mathematics, and philosophy as well as the social systems enabled by these inventions. The ancient Greeks produced and lived in what Lefebvre (1991) calls "absolute" space, which was a "product of consanguinity, soil and language" (48). It is space

> made up of fragments of nature located at sites which were chosen for their intrinsic qualities (cave, mountaintop, spring, river), but whose very consecration ended up by stripping them of their natural characteristics and uniqueness. Thus natural space was soon populated by political forces. Typically, architecture picked a site in nature and transferred it to the political realm by means of a symbolic mediation. . . . At once civil and religious, absolute space thus preserved and incorporated bloodlines, family, unmediated relationships—but it transposed them to the city, to the political state founded on the town. (48)

Absolute space is thus produced, appropriated, occupied, and "owned" by socio-political forces that render it homogeneous, undifferentiated, and ahistorical. Absolute space remained dominant for nearly 2000 years until the technological, political, and economic advances of the early Renaissance produced the second mode of space: "abstract" space. The character of abstract space did not so much destroy absolute space as engulf and incorporate it.

Abstract space is the space of Cartesian coordinates and it accompanies the ascendance of mathematically based knowledges such as science and technology, as well as the emergence of the concept of the autonomous individual. It is the space of wealth and power that "endeavours to mould the spaces it dominates (i.e. peripheral spaces), and it seeks, often by violent means, to reduce the obstacles and resistance it encounters there" (49). Lefebvre explains that it is "abstract" in that it

> functions 'objectally,' as a set of things/signs and their formal relationships: glass and stone, concrete and steel, angles and curves, full and empty. Formal and quantitative, it erases distinctions, as much those which derive from nature and (historical) time as those which originate in the body (age, sex, ethnicity). . . . It has nothing of a "subject" about it, yet it acts like a subject in that it transports and maintains specific social relations, dissolves others and stands opposed to yet others. (49–50)

"Naturalness" is "smashed" and in its place is established "the space of accumulation (the accumulation of all wealth and resources: knowledge, technology, money, precious objects, works of art and symbols)" (49).

One of the distinctive features of abstract space is the extent to which it engenders internal contradictions. Without a grounding in homogeneity, naturalness, consanguinity, and soil, abstract space engenders an arbitrariness to the generation of subjects and social relations and the dissolution of others that inevitably leads to conflict. For example, technologies of mass communication such as the printing press (the effects of which are examined by Eisenstein, 1979) made public access to information widely available for the first time in history. This exerted a potentially democratizing influence in social relations and in the creation of a new (bourgeois) subject. At the same time, however, the accumulation of capital and the subdivision of labor necessary to execute large-scale undertakings such as the printing and the distribution of books gave rise to cultural fragmentation and became the sources of new social inequities, contradictions, and conflicts.

We now live in the third transitional era. This era, like those that have preceded it, is characterized by the momentous cultural recognition that past conceptions of space, whether clearly articulated or not, and the social structures that both underpin and are based in these conceptions, no longer serve as well as they once did. Lefebvre (1991) identifies this recognition at work early in the twentieth century:

> The fact is that around 1910 a certain space was shattered. It was the space of common sense, of knowledge *(savoir)*, of social practice, of political power, a space hitherto enshrined in everyday discourse, just as in abstract thought, as the environment of and channel for communications; the space, too, of classical perspective and geometry, developed from the Renaissance onwards on the basis of the Greek tradition (Euclid, logic) and bodied forth in Western art and

> philosophy, as in the form of the city and town. Such were the shocks and onslaughts suffered by this space that today it retains but a feeble pedagogical reality, and then only with great difficulty, within a conservative educational system. Euclidean and perspectivist space have disappeared as systems of reference, along with other "commonplaces" such as the town, history, paternity . . . and so forth. This was truly a crucial moment. (25)

This moment heralded a deep historical crisis. We are still suffering from the aftershocks of the upheaval, making it difficult to speak to the question of what is to come next. We are still too much in the midst of things. Lefebvre asserts that this crisis, rooted in the irreconcilable contradictions of abstract space, can only be resolved by making the transition to a new space, which he calls "differential" space, the space of plurality and difference. Differential space, Lefebvre maintains, will accentuate difference: "It will also restore unity to what abstract space breaks up—to the functions, elements and moments of social practice. It will put an end to those localizations which shatter the integrity of the individual body, the social body, the corpus of human needs, and the corpus of knowledge" (52). However, Lefebvre's account of the nature of differential space, though generative, is unsatisfying and insufficiently developed. One might presume, however, that it must be neither Euclidean nor perspectivist.

Historical knowledge that relies exclusively on the evidence of inscribed knowledge is insufficient to understand the dynamic and multidimensional character of social space and how it produces, and is produced by, spatial practice. In contrast, we must recognize that history—and cultural knowledge generally—is made not only through the conspicuous efforts of pivotal characters and the occurrence of pivotal events, but through the many and different (and often unseen) ways in which we live everyday life. Words and numbers alone are insufficient to account for all the dimensions of human activity in the making of history. The ways we imagine our own (past, present, and future) roles in social space are unduly constrained by literate and numerate thought; the human imagination, after all, is fundamentally dependent on imagery (Kosslyn & Koenig, 1992).

The connection between "imagination" and "imagery" is more than coincidentally etymological. To systematically direct our imagination, we need a language that is suited to envisioning problems of space. Thinking geometrically suggests the possibility of developing a visual discipline of imaginative thought based directly in images, rather than insisting on the translation of the imagination into words and numbers alone. Such a discipline would be of considerable historical importance. As long as we lack the discipline and the tools to systematically think geometrically on an everyday basis, however, we can hardly expect to envision, much less work toward, anything significantly different than we have now.

Juxtaposing the words "imagination" and "history" may appear to be a contradiction in terms. But history is most often presented to students as a chronological account or summary of pivotal events or key figures. This kind of desiccated presentation is generally no more appealing or interesting to most audiences than is the standard presentation of geometry. The inevitable result is that history is likewise a rote subject, devoid of imagination and excitement, as well as completely divorced from our everyday lives and amounting to nothing more than a set of facts to be learned and then forgotten as quickly as possible.

At a more advanced level, history might be said to be about the problem of what "really" happened; this is where the study of history begins to get interesting. Unfortunately, by the time most students get to this point they have long since turned off their imaginations in favor of the disciplined and analytical nuances of professional and scholarly discourse, which are directed primarily toward the sort of knowledge that can be contained in linguistic and mathematical expression. As long as we continue to rely on words and numbers as our primary (and pedagogical) tools for the construction, communication, and interpretation of history (and knowledge generally) we will never be able to imagine, much less construct, more than a limited, partial and biased account of the multiple dimensions of human experience, especially those dimensions that resist expression in words and numbers.

Contemporary systems of education consistently tend to overlook the spatial and multidimensional character of knowledge. This oversight is based in a contradiction that ultimately undermines the very effectiveness of those systems. The word "education" has its etymological roots in the Latin *educare*, literally, "drawing out" (*Websters*, 1983:576). More often than not, however, students get "drawn *on*"—in the sense both of being pulled or pushed along, and of being labeled and graded through the organized discipline of educational institutions. Students are, more often than not, trained in limited specializations, rather than led to develop their own minds and imaginations. The key role of drawing, geometry, and geometrical thinking in the "drawing out" of knowledge remains unrecognized and untapped as long as reading, 'riting and 'rithmetic are the *only* tools one has for such construction.

Through the exercise of the geometrical and historical (spatial) imagination, it may be possible to arrive at a new understanding of the relation between words, numbers, and pictures as collaborators rather than competitors in the production of knowledge. We all, individually and collectively, produce—that is, construct—not only meaning, but the character of social space, history, and knowledge generally. We accomplish this not only through inscriptions but also through the manifold trajectories of our own lives. To the extent that one ignores this knowledge, remains unaware of it, or fails to take advantage of it, one abdicates the production of both space and history to others whose self-interest and goals may be contrary to one's own. Thinking geometrically is a way of asserting one's own interest in the (social) production of (social) space. What this means in practical terms might best be understood by taking a look at where the phrase "thinking geometrically" came from in the first place.

Geometry in Action

Explicit references to geometry in popular literature are so few and far between that they really stand out. One example in particular has remained with me for more than a decade: Robert Ludlum's *The Bourne Supremacy* (1986), an action novel from the latter days of the Cold War. This book, like any other fictional work, might be considered an imaginary history. Its key characters may not have exact parallels in the real world, but no doubt they have either partial or composite equivalents. It isn't too difficult to imagine that behind the scenes, unreported in the news, people carry on activities much like those undertaken by the characters invented by Ludlum. One might consider *The Bourne Supremacy* a parable for its times: the era of late Cold War tensions, an era which today seems quite remote and is recognized as having been as much a product of our imaginations as any work of fiction. But the ways in which Ludlum's hero resolves questions of uncertainty and identity through the exercise of geometrical thinking and action set an example for contemporary life.

The Bourne Supremacy is the second novel in a series—the Bourne trilogy—in which the action revolves around a chameleonic protagonist who alternates between multiple personalities and identities:

- David Webb: A promising young diplomat, whose wife and children are killed in a mistaken air raid in Indochina. After an extended period of revengeful insanity, he remarries and becomes a professor of Asian studies at a small Northeastern college.
- Delta: Webb's earliest reincarnation as a suicidal yet ever-prevailing Vietnam-era commando, who belongs to an outfit code-named "Medusa." He is driven to desperation by a need for revenge on his family's unknown killers.
- Jason Bourne: Webb's next life as a CIA agent who snaps and goes "off the wire"—out of control—in clandestine operations, builds a reputation as a mad killer for hire, and manages to "lose" millions of dollars in agency funds. He is labeled beyond salvage and is nearly killed by his employers when he tries to "come in from the cold" after recovering some of his memory and sanity.

Ludlum's choice of names has considerable mythological and historical resonance. "David," in the Old Testament, slew Goliath. "Webb" evokes the work of Ariadne and is a play not only on the multiple strands of the central character but on the complex network of events he both weaves and becomes caught up in, much as the reader is caught by the tale spun by the author Ludlum. "Delta," the fourth letter of the Greek alphabet—characterized by a triangle—is today used as a mathematical symbol for change or difference. Delta's outfit was named after the snake-haired Gorgon Medusa, whose terrible gaze turned men to stone even after Perseus beheaded her; they, similarly, were a treacherous bunch of snakes, many of whose deeds were deliberately aimed at paralyzing opponents with fear. Finally, "Jason" is a reborn version of the original Argonaut—the archetypal antihero, sent against his own will on a quest for the Golden Fleece. Lawall (cited by Simpson, 1976:64) describes the ancient Jason as a "symbol of the unique predicament and responses of Hellenistic man. . . . Traditional values of heroism, honor, and integrity are jettisoned as Jason puts in motion the machinery of success." The Jason of Ludlum's story is in many ways symbolic of the unique predicament and responses of Western technological man. What distinguishes the modern Jason is how he ultimately recovers these traditional values even as he turns them inside out in the process. The modern quest of Bourne/Delta/Webb can teach us something about thinking geometrically.

In the story of *The Bourne Supremacy*, David Webb has recovered some of his memory and has found peace of mind with his second wife, Marie, an economist who helped him appropriate and launder the "lost" millions as recompense for his mistreatment. Her character, and even her name, to stretch things a little, parallel the mythological Argonaut Jason's first wife, Medea, who was instrumental in capturing the Golden Fleece. David Webb, trying to put his other identities to rest, has settled into a quiet career as a college professor. Suddenly, he receives several mysterious phone calls and rushes home to find that Marie has been kidnapped. Through a reversion to the ruthless investigative capabilities of Delta and Bourne, Webb figures out that his former CIA handlers—the analyst Edward McAllister and the "ambassador" Raymond Havilland—have, for reasons unknown, arranged Marie's kidnapping to maneuver him into becoming Bourne again. All Webb really knows for sure is that in order to recover her and, hopefully, the stable life they had together, he must go to Asia and track down a ruthless killer who looks like him, acts like him, and even claims to *be* Jason Bourne. This impostor's continued activity has everyone but Webb, McAllister, and Havilland convinced that the real Jason Bourne is back in action. It is Webb's—or, rather, Delta and Bourne's—job to set things straight, and he—they—do so with a vengeance.

Bourne/Delta/Webb discovers that the impostor was trained by Philippe d'Anjou, a former colleague in Medusa. D'Anjou, recognizing the monster he has created, turns against his renegade protégé and helps Bourne/Delta put the impostor out of action, a reversal for which d'Anjou pays with his life.

When the real Bourne finally captures his false double, Bourne realizes that they differ in one significant respect:

> The fact that the looped rope would not have choked the impostor, any more than a loose clothesline wound around a person's neck would, told Delta [-Bourne] something. His prisoner could not think geometrically; stresses were not a strong point in the killer's imagination, only the spoken threat of death. It was something to bear in mind. (451)

In confronting the false Bourne with the difference, Delta /Bourne draws out further this concept of thinking geometrically:

> "Think of d'Anjou," said Jason flatly.
> "He *wanted* to die, for Christ's sake! What the hell was *I* supposed to do?"
> "Nothing. Because you [an impostor] are nothing."
> "Well then, that kind of puts me on *your* level, doesn't it, sport? [D'Anjou] made me into *you!*"
> "You don't have the talent," said Jason Bourne. "You're lacking. You can't think geometrically."
> "What does *that* mean?"
> "Ponder it." Delta [Bourne] rose to his feet. "Get up," he commanded. (452–453)

The reader, too, is left to ponder the notion of "thinking geometrically." Ludlum never explicitly develops what is meant by this apparently offhand remark, although he repeats, or more specifically, recapitulates it several times in the course of subsequent action. Thinking geometrically is clearly the difference between the real and the false Bourne, and the reason that the real Bourne prevails. We, like the impostor, are left to develop our own understanding of what thinking geometrically means.

I can best express my own understanding of what Ludlum means in abstract terms. It is the ability to visualize events and actors in multiple dimensions and various possible planes of action. These planes or dimensions constitute a complex space of possibility. Because one is limited by the ability to apprehend only a few dimensions and planes at once, he or she who can manage more of them than the competition can will enjoy a considerable advantage.

Thinking geometrically means not so much the ability to "keep things in perspective" as it does the ability to perceive and manipulate the transformations among different points of view as well as possible planes of representation and action. In the above excerpt, this is shown by the alternating identities and perspectives of Delta, Bourne, and Webb. Bourne shifts effortlessly into the persona of Delta once the situational exigencies require a different mode of action. The trinity of Webb/Delta/Bourne is not quite three-in-one; it is a nesting of dualities—primarily, Delta alternating with Bourne. Webb is in a sense in opposition to both of them, but he must use them for his own and Marie's survival. The situational plane of action determines who acts and how at any moment. Webb could no more see or live in Delta or Bourne's world than they could in Webb's. Yet each is crucial to the action. Without Webb there would be no need to act. Without Delta/Bourne there would be no action.

The action proceeds. Delta/Bourne, thinking geometrically all the while, proceeds to use the false Bourne in a plan to recover his—or, more exactly, Webb's—wife unharmed and then to exact vengeance on McAllister and Havilland. In executing this plan, he appears to have gone "off the wire" again. That is, he has crossed the thin line between commando and terrorist, from being sanctioned, however unofficially, by the government, to acting in his own interest, an unsanctioned cause.

Delta's exponentially vengeful rage is finally contained by the intervention of Marie, who has escaped her kidnappers but now collaborates with them out of desperate concern for the safety of her husband. With the help of Alex Conklin, a former CIA friend of Webb's, she turns to McAllister and Havilland—the ones who set up her kidnapping—to save her husband both from himself and them. Her intervention during a violent and climactic confrontation between Bourne/Delta/Webb, McAllister, and Havilland at a Hong Kong safe house forces the out-of-control Bourne/Delta out of the current plane of action and brings Webb back into control of himself, but also back under control of his (and her) handlers: McAllister and Havilland. They, perhaps even more than he, are conscious masters of the art of thinking geometrically; they take it to another, wholly impersonal, and possibly inhuman level, thinking globally but at a distance from the fray.

The scope of their vision and the depths of their imagination become gradually apparent when Webb, in an angry confrontation with Havilland and McAllister, learns that bringing in the false Bourne was only a first step. The real target is someone far more dangerous: Sheng, a ruthless Chinese politician aspiring to wrest Hong Kong from British rule as a step toward gaining control over all of China, a prospect that threatens to undermine the stability of existing global power relations. The false Bourne was only one of Sheng's many agents, and the true project of stopping Sheng is unfinished. Webb realizes that he, as the real Jason Bourne, is uniquely qualified to finish the job. He agrees to go back under cover to do the rest when he recognizes the scope of the situation.

That Bourne/Delta/Webb has again been manipulated, but now by the truth (or at least, a part of it) rather than by lies, is made chillingly clear in a subsequent discussion between Havilland and his friend Alexander Conklin:

> "Oh, *Jesus!*" whispered Alex [Conklin], mesmerized by the cold, precise elder statesman of covert operations. "You told [Webb] everything because you had your own backups. Even if he was taken, you could cover your ass with official logs, psychiatric evaluation—you could disassociate yourself! Oh, God, you *bastard*."
>
> "I told him the truth because he would have known it if I tried to lie to him again. McAllister, of course, went further, emphasizing the organized crime factor which is all too true but a sensitive issue I'd prefer not to bring up. Nobody does. But then I didn't tell [McAllister] everything. He hasn't yet put enough distance between his ethics and the demands of his job. When he does, he may join me on the heights, but I don't think he's capable."
>
> "You told David [Webb] everything in case he was taken," went on Conklin, not listening to Havilland. "If the kill doesn't happen, you want him taken. . . . Then Sheng will get the message that his conspiracy's known to us and he'll get it *unofficially*, not from us but from an unsanctioned mental case. *Jesus!* It's a variation of what Webb *told* you!"
>
> "Unofficially," agreed the diplomat. "So much is achieved that way. No confrontations, very smooth. Very cheap. No cost at all, really."
>
> "Except a man's *life!*" shouted Alex. "He'll be *killed*. He *has* to be killed from everyone's point of view . . ."
>
> "There are more important priorities here than the survival of a single individual, Mr. Conklin, and again you know that as well as I do. It's been your life's work—if you'll forgive me—but the present case is on a higher level than anything you ever experienced. Let's call it a geopolitical level." (532–534)

Of course, Bourne/Delta/Webb finishes the job and comes back in safely, working with McAllister (whom Havilland thought wasn't "capable," ironic words that parallel Bourne's assessment of the impostor: "you don't have the talent"). There are happy endings on several levels. McAllister conquers his own limitations, perhaps by osmotic association with Bourne, and learns in his own way to think geometrically despite his ethical conflicts. The Webbs are reunited, hopefully for good. The stability of existing global power relations, though momentarily threatened by a series of eruptions both generated and contained by Bourne/Delta/Webb and his handlers, is ultimately restored, much as the Egyptians restored field boundaries after the seasonal floods of the Nile receded. It's a ripping good yarn, and one in which the surface action tends to disturb distinctions between good and evil, truth and lies, on planes of action from the local to the global.

The disturbances wrought by the character and actions of both the modern and the mythological Jason compel us to wonder: Is it ever possible to distinguish between such apparently fundamental categories as good and evil, truth and lie? For our current purpose, it is sufficient to note that among the myriad possible answers to these questions, ambiguities and distinctions are all too often ignored or overlooked in the face of exigency. The theoretical prospect of affecting the quality of our lives—making history—adversely or otherwise through one's choice of action is often outweighed by the immediate and practical need to save one's skin when facing the threat of death. Variable dimensions and planes of action matter.

Like the Argonaut Jason, Bourne is on a quest that he has been maneuvered into by forces beyond his control, and both are perhaps less heroic than they are antiheroic in their pursuit of success in their respective adventures. In the modern story, the "fleece" turns out to be anything but golden. It is a sham, or perhaps better, a scam: not an object or even an idea, but a process, a multidimensional con game, played for keeps at every turn. Every participant-artist in it can only save his or her own skin by fleecing others. In this sense, neither Jason is a suitable model for everyday practice; no one in their right mind would imitate either of their lives.

More important, Ludlum's story is of interest because of the way in which thinking geometrically is identified as a primary attribute in constructing and navigating successfully multiple, illusory, treacherous, and violent environments. In fact, part of the treachery and violence of these environments is rooted in their being perceived as two-dimensional, partial representations of a multidimensional world. Without some means of reconciling the contradictions and illusions of such limited representations, no appropriate response to the conditions they present is possible.

The advantage Jason Bourne has in this regard, and which he exploits to the utmost, is thinking geometrically. Bourne is a multiple character, a chameleon who changes to meet different conditions. So are the other players, but they just can't keep up with him. He manages to think and act along more dimensions of their collective space than anyone else. Taken together, the trinary character of Bourne/Delta/Webb demonstrates, often simultaneously, several distinct dimensions of experience and action: pattern visualization (rehearsing actions in the mind's eye); spatial reasoning (understanding stresses, actions, and reactions viscerally and intellectually—like an engineer at home with machinery); physical coordination and kinesthetic awareness (understanding one's involvement in the flow of events: as Wayne Gretzky put it, knowing to go not to where the puck is now, but to where the puck is going to be); design ability (planning for and balancing among multiple evolving contingencies); abstract reasoning (understanding hierarchical events and causality); and, of course, verbal or rhetorical ability (being able to persuade or manipulate others whose involvement is necessary).

Thinking geometrically is not necessarily logical or intuitive; perhaps it is neither or it is both. It cannot be reduced to any single component or representation of the whole. It is the ability to hold together a larger picture in one's mind while partial views only are available to the senses, and to act in accordance. The multiple persona of Bourne/Delta/Webb does this intuitively or subconsciously through the simultaneous exercise of both serial and parallel thought processes and actions.

This is rather like being both airport traffic controller and pilot of an airplane at the same time. For efficiency and safety, such roles are rigidly separated under normal conditions; in Ludlum's nether world of both sanctioned and unsanctioned political intervention, shifting conditions on parallel and simultaneous planes of action rarely allow the luxury of such a division of labor. Bourne/Delta/Webb must divide and assemble himself appropriately to address simultaneous and multiple possibilities and contingencies at any given moment. The spoken or unspoken threat of death is always just around the corner.

Thinking geometrically means not only being able to meet any situation with the resources at hand, but to realize that one's involvement in it creates a new situation. It also means being able to take one's thinking to a place apart from the current situation while remaining involved. It requires both an analytic and a synthetic perspective. Neither perspective taken alone is as powerful. On the one hand, analysis requires the distancing, or uninvolvement, of the observer from the observed; it is a powerful tool under the right conditions. It is perhaps the dominant mode in print-based culture. Analysis, in the form of subdivision or making distinctions, can be carried out indefinitely which is, as Foucault (1973:9) observed regarding the relation of language to painting, an "infinite relation." Synthesis, on the other hand, draws things together and tends to blur distinctions. In so doing, it appears to lead one into an endless field of ambiguous gray instead of precise black and white; it threatens us with entropy, a homogeneous state where further synthesis is meaningless. It carries with it a hint of the negative: "synthetic" is a modern synonym for "not the real thing." Not surprisingly, synthetic thinking is a rare trait in Western culture. We have all manner of job titles with the word "analyst" in them, but none I can think of with the word "synthesist." Either analysis or synthesis, taken to the extreme, leads to infinity, the lonely "heights" (or depths, depending on your point of view) inhabited by the likes of Havilland.

It's no accident that when Delta/Bourne returns to normal (literally, a word meaning perpendicular, or in his case, perhaps "square"), he is David Webb, a college professor and a stereotypically analytic character. His ability to think geometrically is apparently a temporary aberration, a protracted episode of mania that he recoils from when Delta/Bourne has succeeded. It is testimony to the integrity of his analytic and synthetic capabilities that once he sees who the false Bourne is acting for—the ruthless and ambitious Sheng—Webb voluntarily returns to the persona of Delta/Bourne to fully resolve the looming global/historical/political crisis that he now understands through the combined experience of his selves. Webb becomes the center of revolution (both geometrically and politically), in control of not only himself (for the first time) but of the entire world. His only real recompense is the promise that both Delta and Jason Bourne will be finally put to rest and that he, as David Webb, will be allowed to live in peace.

The conventional conclusion one might draw from this is that Bourne's action-oriented, geometrical thinking is generally at odds with the status quo. In an earlier Ludlum novel, *The Bourne Identity* (1980), the uncontrolled existence of the geometrical thinker/actor Delta/Bourne/Webb represented such a threat that he was nearly killed when he appeared to have gone "off the wire" for the first time. After the situation was finally sorted out he was put into supposedly permanent retirement. He was recalled only when other circumstances become even more threatening.

Is it appropriate, or carelessly subversive, to suggest that such a dangerous mode of action as thinking geometrically ought to be cultivated deliberately? The association of geometrical thinking with physical action and violence, as demonstrated by Ludlum's story, suggests that cultivating it might be a formula for cultural disaster. In a stable world, this might be the case; in a dynamic and unstable world, however, it might be the only solution.

The challenge of thinking geometrically is formidable. Although we may lack formal systems for designing and communicating the multidimensionality of objects, processes, and events in the real world, we still manage, with varying levels of success, to cope with the multidimensional unpredictability and contingency of the world through action. Action, uninformed by theory, can be little more than either habit or trial and error. Learning to think geometrically is a step toward understanding and directing better our own actions, by transcending the artificial boundaries between words, numbers, and pictures, and ultimately discarding the counterproductive habits we derive from our limited experience in these domains. Along with a greater awareness of the complex mechanics of the world comes an awareness of how suddenly an apparently stable order can come tumbling down. Any of us, like David Webb, can be either an agent or a victim of change, and possibly both, but nobody wants to bring things down on one's own head. One who has power, and uses it, has to have a sense of responsibility: to know when to stand still, when to tread carefully, and when to act forcefully.

The fictional example of Jason Bourne has served to demonstrate the practical origins and applications of thinking geometrically in the imaginary (that is, fictional) making of history. It is also an example of the risks associated with the exercise of thinking geometrically. For a nonfictional historical (and more theoretical) demonstration of how thinking geometrically served to change the world, I turn next to the example of Galileo Galilei.

The Mechanics of History

In *Man and Crisis*, Ortega y Gasset (1958) predicts a flowering of the historical imagination on the order of the scientific imagination unleashed by Galileo.[2] Ortega y Gasset writes:

> This is the way Galileo begins the fourth chapter of his last book entitled *Dialogue of the New Sciences* . . .
>
> "I conceive as the work of my own mind a moving object launched above a horizontal plane and freed of all impediment." That is to say, an imaginary moving object on a plane which is ideally horizontal and free from any hindrance whatsoever—but those hindrances, those impedimenta from which Galileo imaginatively frees his moving objects are the facts, for every observable body moves in the midst of impedimenta, knocking against other bodies and by other bodies knocked. So he begins by constructing a reality, doing it mentally and ideally. Only when he has his imaginary reality well in hand does he note the facts, or rather, does he observe what relationship obtains between the facts and the imagined reality.
>
> Having that background in mind, I hold the conviction that we are now approaching a splendid flowering of the historic sciences, thanks to the fact that historians will resolve to confront historical facts just as, mutatis mutandis, Galileo confronted physical facts. They will become convinced that science—by which I mean the entire body of knowledge about things, whether corporeal or spiritual—is as much a work of imagination as it is of observation, and that the latter is not possible without the former; in short, that science is a process of construction.
>
> The imaginative character of science, imaginative at least in part, makes it a sister of poetry. But between Galileo's imagination and that of a poet lies a fundamental difference—the Galilean imagination is exact and precise. Both the moving object and the horizontal plane which his mind conceived are figures that are rigorously mathematical. But historical material has basically nothing to do with mathematics. Must it therefore renounce all thought of being a construction, which is to say a science, and declare itself to be nothing but poetry? Or is there room for a type of imagination, which, without being mathematical, will lend to history the same strictly constructive service which mechanics renders to physics? Is there room for a quasi mechanics of history? (15)

Four decades have passed since these optimistic words were written, but we have yet to see the "historic sciences" emerge from the shadow of the natural and physical sciences, which have flowered bountifully during the same interval. One must wonder: What happened? Could Ortega y Gasset have been so far off the mark?

Rather than answer these questions directly, I would rather call attention to the fact that in the quotation above, Ortega y Gasset tempers his prediction with several questions. Note that he does not ask whether there *is* a type of imagination that will serve history as constructively as mechanics serves physics; he asks instead, twice, whether there is *room* for such an imagination, and this is the heart of the matter.

The work of the imagination, like that of Jason Bourne, is essentially covert. No one can see what goes on in the brain; the only real evidence of imagination we have is what someone produces or does as a result of thought, imaginative or otherwise. Documentary evidence, such as words, numbers, and pictures, is what we most often rely on to prove the existence of the human imagination. Working backward from this evidence to speculate on the workings of the imagination is itself an exercise of the imagination. To understand the imagination, the observation of the practices involved in the production of this documentary evidence may be better yet, since they are closer to the source. The study of such practices may tell us much more than static documents do about the dynamic character of the imagination and how it works. However, sometimes the documentary record is all one has to work with, as is the case with Galileo. Ortega y Gasset describes the problem Galileo faced:

> If motion has no structure which is essential and always the same, and of which the individual movements of individual bodies are mere variations and modifications, then a science of physics is impossible. . . . It is necessary that in the movement of smoke rising from a village chimney and of a stone falling from a tower there be, though seemingly opposite, an identical reality; that is, that the rising and the falling obey the same laws. (18)

After the fact of Galileo's discovery of the law of falling bodies, it is easy to overlook the fact that arriving at it required the exercise of a new kind of imagination. A hint of the character of this imagination is contained in the encyclopedic summary: "Perhaps the most far-reaching of Galileo's achievements was his re-establishment of mathematical rationalism against Aristotle's logico-verbal approach, [through which] he was able to found the modern experimental method" (*Encyclopedia Britannica*, 9:1088).[3] Galileo had to have consciously recognized that this new rationalism could not be spoken or written through ordinary language when he asserted that "the Book of Nature is written in mathematical characters" (*Encyclopedia Britannica*, 9:1088).

The importance of modern mathematics in the emergence of science cannot be overstated. Bruno Latour (1990), in examining how contemporary science came about, observed that "no scientific discipline exists without first inventing a visual and written *language* which allows it to break with its confusing past" (36; emphasis added). Prehistoric and oral culture was rife with contradictions and limitations that could only be fully understood (and repudiated) by the invention and application of a better language. Aristotelian physics, based in the once new form of "logicoverbal" thought enabled by the invention of script allowed philosophy to make that break. The dominant language of modern physics (after Galileo) was, and still is, mathematics.

In science, the validity of experimental results is established through their replicability. That is, another experimenter ought to be able to repeat the experiment and arrive at the same results. However, it's one thing to replicate any of Galileo's experiments and another to replicate the work of his imagination. Neither science, nor history, nor any other existing discipline serves very well to guide one in such an effort; if anything, these disciplines present obstacles. Established systems of knowledge and belief not only structure one's thought and imagination along the lines of what is already known; they also inhibit the investigation of alternative systems. To find out more about the imagination, we must look between the lines of the documentary record.

The conspicuously mathematical character of Galileo's work makes it easy to overlook the important role of a disciplined sense of vision in his discoveries. What Galileo demonstrated was more than just an imaginative solution to a difficult scientific problem; he worked out a new geometric conception of space in which his imagination, like an imaginary projectile, found room to move, freed of all impediment.[4] Latour (1990) asserts that the geometric art and science of *perspective* is more than anything else what made this new space possible: "The rationalization that took place during the so-called 'scientific revolution' is not that of the mind, of the eye, of philosophy, but of *sight*." He asks, "Why is perspective such an important invention?" Quoting Ivins, he answers: "Because of its logical recognition of internal invariances through all the transformations produced by changes in spatial location." He continues, "Perspective . . . is an essential determinant of science because it creates 'optical consistency,' or in simpler terms, a regular avenue through space" (27). Within this space, Galileo found a mathematical solution to reconcile the apparently contradictory movements of different bodies into a unified, consistent and logical scheme of motion. Without the "optical consistency" provided by perspective, this solution may well have been impossible to see.

Perspective, invented by Brunelleschi in the 1500s, is more than a set of systematic techniques for graphic representation; perspective serves also as a model for thought (Damisch, 1994). Even though it would be difficult to sustain any claim that perspective itself constitutes a "language," it has nonetheless served to shape both language and mathematics. For example, equitone prose, which McLuhan (1962) calls the literary analogue of graphic perspective, is the standard for modern writing. In science, the model of perspective served not only to liberate the spatial imagination (especially after Descartes' invention of coordinate space[5]) by locating the experimental observer outside of the particular and local time and space of the events or objects observed; it also served theoretically to remove the "impedimenta" of that time and place. In science, writing is also deeply perspectival (Bastide, 1990), but it is intended to *constrain* the reader and his or her conclusions. The modern scientific essay "is constructed like some military strategy, an ambush without an escape route. Each time that a reading of the results different from that of the authors could possibly be made, the departure is barred by an adequate argument" (Latour, 1990:208). Just as all roads once led to Rome, in both modern science and literature all lines of thought converge to a central "point of view" established through perspective.

Although I return to a consideration of perspective in chapter 4, it is important to recognize here that perspective provided the basis for a new and rigorous kind of spatial imagination. Latour attempts to describe how Galileo's imagination actually worked through (and with) the disciplined sense of vision made possible by perspective:

> A simple change in the geometry used by Galileo allows him to connect many different problems, whereas his two predecessors worked on disconnected shapes over which they had no control. . . . In fact, what gets connected are three different visual horizons held optically because the surface of the paper is considered as geometrical space. (48)

Galileo's imaginative work is a prototypical example of thinking geometrically. He is able to connect imaginatively multiple visual horizons—both across and along time—through the use of graphic images, which are static representations of either simultaneous or serial events, and through direct observation. He is also able to control, connect, and compare this evidence with both imaginary events and physically observed events. Through the imaginative effort of thinking geometrically, perhaps anyone, like Galileo, might discover patterns of behavior or events that enable one, with the aid of a suitable language, to explain or predict as well as to act.

As yet, history lacks a suitable language for this kind of imaginative work. Words, numbers, and pictures are the primary evidence we must rely on to investigate geometrical thinking. However, our reliance on language and mathematics as disciplines of knowledge also inhibits our understanding of the imagination. Taking Galileo as a model, thinking geometrically may depend most of all on the sense of vision. We still don't really know much about pictures or how pictures relate to words and numbers. The geometrical art and science of perspective, the invention of which was crucial in establishing Galileo's new kind of science, once served to liberate the spatial and geometrical imagination and held the promise of being a language of vision, but this promise dissipated as perspective was translated into its literate and numerate forms. In these forms, perspective is as entrenched and pervasive a model as Aristotelian logicoverbal thought was in Galileo's day.

The very factors, once so liberating in Galileo's thought, now impede the development of the geometrical imagination. Today, what Gardner (1983/1993, 1993) calls "logico-mathematical" and "language-based"[6] modes of thought are dominant throughout our culture and leave little room for other modes of thought. Other modes (Gardner lists seven "intelligences"[7]) are simply crowded out of the picture, or subordinated to these two dominant modes. For example, even though the faculty of vision is a practical necessity for reading and writing either linguistic or mathematical texts, the specialized (and perspectival) disciplining of the visual sense achieved through literacy and numeracy tends to limit any extension of the sense of vision through other forms of imagination and inhibit the development of a "quasi mechanics of history" that might be based on such an extension.

These limits can be seen in both theory and practice. As geographer Edward Soja (1989) observes, despite "the critical and potentially emancipatory value of the historical imagination" and its centrality in critical social theory (14), the historical imagination all too often lapses into an extreme form: "historicism." Soja defines historicism as "an overdeveloped historical contextualization of social life and social theory" that "actively submerges and peripheralizes the geographical and spatial imagination" (15). This contextualization is achieved primarily through language, and in particular, writing. After all, there is no "context" without texts in the first place. However, the historical imagination is poorly served by its dependence on spoken and written language, which, though adequate in accounting for situations that can be reduced to linear, logical, temporal, or narrative sequences, is not well suited to account for more complex and fundamentally spatial conditions such as simultaneity—parallel events—or for reconciling multiple points of view. Soja, drawing on the work of Lefebvre, sees such problems as essentially spatial in character and asserts, therefore, that as a discipline history ought to be deeply concerned with problems of space. Without a suitable language, history and the historical imagination remain mired in a "confusing past." Just as those who are ignorant of history are doomed to repeat it, historians who ignore the spatial character of history—and how their work projects onto the world and onto history itself—inevitably spiral into the trap of historicism.

The space of contemporary history is a limited and at most two-dimensional space suffering from two particular restrictions. First, a historical point of view is always located in the present and directed toward the past. Second, history is always written. Oral or pictorial historical evidence is generally handled as either unreliable or ambiguous and validated only through translation into textual form. To the extent that history is the business of finding key principles and events among the welter of evidence in order to account for what "really" happened, and the extent to which it relies solely on the perspective of language, its efforts will continue to be counterproductive. For history to get into the business of imagining both the present and the future as well as our relationship with the past, we must find better tools.

Language, like mathematics, is but one of many possible models for thought and representation. Despite the power and flexibility of words and numbers, not everything is successfully reduced to either. It is not too difficult to recognize that mathematics is an unsuitable tool or model for the historical (and spatial) imagination, but it is harder to recognize, or accept, the possibility that language may be equally unsuitable. History attempts to define and locate the causes and effects of actions or events that change the parameters of human knowledge and potential. It is constructively opposed to mythology, which explains the responsibility for such actions as located outside the realm of human activity: through the intervention of gods, demigods, or demons. However, the heavy reliance of history on perspective—especially in its verbal and mathematical forms—as a model for the imagination and construction of knowledge limits its own ultimate prospects for explaining anything at all.

At least graphic perspective is visually explicit and its distortions more recognizable. In verbal or mathematical perspective the inevitable distortions are implicit: they are not so visible or readily detectable. The shared subject matter that is common to even slightly different points of view may be concealed by their more obvious differences, which may appear to be irreconcilable from each (necessarily limited) perspective. Perspective generally fails when presented with multiple, contradictory views of what "really" happened in a given situation, sometimes referred to as the Rashomon syndrome.[8] Common sense and experience tell us that there is not only a multiplicity of motives and interests behind any human event or action, from holiday celebrations to armed conflicts, but that there is also always more than one possible point of view from which to interpret them.

What neither common sense nor history can do is reconcile these multiple causes and differences within a consistent theoretical framework that enables any kind of accurate explanation or prediction of what appear to be mutually exclusive accounts of either the same or different events. Science promises us such a capability in regard to events in the world of physics. Recall Galileo's statement that "the rising and the falling observe the same laws." As yet, however, science and history are of limited use in accounting for the past, understanding the present, or predicting the future in the world of human affairs.

One of the key lessons to be learned from modern mathematics and science, which historians might do well to learn, is that even though causes may not be explicable, knowledge can still be produced (Kline, 1985). For example, even though Galileo's laws predict the movement of mundane bodies in space, these same laws tell us nothing about why they behave the way they do or what causes their movement in the first place.[9] However, this lack of explanatory power has no effect at all on the validity of Galileo's laws, which work whether we understand them or not.

The contemporary historical imagination is unduly constrained by a concern with causes and a primary dependence on the limited (because perspectival) tools of language—spoken, written, and to a certain extent, mathematical (statistics, at least)—to account for these causes. It is impossible to deny the fundamental importance of language and mathematics in the construction of knowledge. In fact, written language is here my own primary tool. But it may be necessary to discard both the notion that causes ought, or need, to be explicable and the notion that words, numbers, and pictures are up to the task in the first place.

Perspective was crucial in establishing not only a consistent and logical frame of reference—a space—but also a means—a visual language—for representing objects and their relations in the abstract space that, according to Lefebvre, emerged in the early Renaissance and persists today. More than providing a means for the "optically consistent" and logical representation of space, perspective also enables the consistent and logical expression of this space in linguistic and mathematical terms, thus, and ironically, serving to underscore the priority of language and mathematics in the construction of knowledge. Graphic perspective, like verbal and mathematical texts, require a two-dimensional discipline of vision, but it is inherently limited and too bound up with language and mathematics to serve as a language of space. It is perspective assimilated into language in the form of literacy that today shapes our thought more than does graphic perspective. We tend not to trust that which can only be shown in pictures. However, we cannot afford to simply ignore that which remains outside the sphere of language.

Language itself must be understood as untrustworthy. Henri Lefebvre (1991) asserts that languages, as we know and use them, are inadequate for either the identification or the solution of problems of space:

> Languages, each of them in particular and all of them in general, all linguistic systems, including that of established knowledge, are spoken and written in a time and space to which that knowledge tends to assign a privileged metaphysical status. They are clumsy in the way they give utterance to social time, to spatial practice. . . . The fact is that languages and linguistic systems need to be dismantled and reconstructed. This task will be carried out by and in (spatial) social practice. (414)

With either spoken or written language, much is tied up with one's choice of working terminology; the words one uses are limiting factors indicative of the assumptions one has made regarding potential conclusions. Therefore, a simple translation may be critical in addressing Ortega y Gasset's questions, here substituting *space* for *room:* Is there space for a type of imagination, which, without being mathematical, will lend to history the same strictly constructive service that mechanics renders to physics? Is there space for a quasi mechanics of history? This small semantic change relocates his questions into a realm of more complex dimensionality. Historical problems might be fruitfully viewed as problems of space, amenable to investigation through the efforts of a rigorous and precise, yet imaginative and geometrically inclined, mind.

The obvious—but difficult to precisely define or explain—differences between words, numbers, and pictures are irreconcilable within the two-dimensional framework of perspective. However, through a sort of Galilean move of constructing a new concept of space and developing a suitable language to consistently and rigorously connect these different modes of communication (each of which is a limited form of two-dimensional representation), it should be possible to arrive at a more comprehensive understanding of the space that they collectively inscribe.

If, as Lefebvre asserts, space is *produced* through human activity and effort, the study of geometry theory and graphic practice provides a way of opening up a quasimechanical yet nonperspectival understanding of history as well as other social and cultural phenomena. This might eventually lead us to a flowering of the historical sciences such as the one we have witnessed in the natural and physical sciences over the last several hundred years. Through the development of a quasimechanics of history we might be able to account as precisely for human behavior or the behavior of abstract bodies such as corporations or governments as we do for the movements of falling bodies or other mechanical objects like planes, trains, or automobiles. Without such a science, the traces that nonmechanical entities leave in time and space must remain as mysterious as the movements of celestial bodies must have been to cultures without an astronomy.

Were such a scientific and historical discipline to exist, one might say that its practitioners—whether theoretical, like Galileo, or empirical, like Jason Bourne—would themselves be the mechanics, or perhaps better, the engineers of the past, the present, and the future. Although a science of history does not yet exist, it is not too difficult to imagine that it could. What is more difficult to imagine is the effects that such a development might have on culture generally. But some insight might be gained through the study of past changes in human knowledge, such as were once inspired by the invention of perspective.

Galileo's work, whether he intended it or not, served to transform culture in revolutionary ways. In his own time, this effect might not have been fully recognized. After all, God was the Great Designer and Man was but the executor of His divine plans. But in retrospect, we can see that knowledge and culture are constructed through human activity and intervention in the world. By and large, we no longer either credit or blame God for whatever happens, for better or worse. We are, each and every one of us, and all together, the architects, engineers, builders and maintenance mechanics of contemporary culture and of our future. To make the most of this role, rather than abdicate it through ignorance or negligence, we must develop a geometrical understanding of the spatial character of the forces and relations (not all of them mechanical in nature) that constitute what Lefebvre would call the present mode of space in which we live. Unless we are willing to trust entirely in Providence to solve the social and environmental problems that confront us, we must, each of us and collectively, take on that responsibility. Doing the job well, while recognizing the consequences and stakes involved, requires substantial effort of imagination at many different levels. Ultimately, the historical imagination will not be seen to be much different from the geometrical imagination, and the language that best connects their workings might be a fundamentally graphic one.

The project of dismantling and reconstructing language, as Lefebvre suggests, is but one step in the process of establishing a new relation of graphics to words and numbers. Developing a quasi mechanics of history might be a long-term consequence of such a project. But it is more immediately important to be concerned with the fact that within the perspectival framework of established bodies of knowledge—that is, between the cultural and disciplinary boundaries such as those maintained between poetry and mathematics, or history and science—there is little room for a flowering of the historical or spatial imagination. In fact, the ascendance of science and technology in contemporary culture presents us with the prospect that the historical sciences may be on the verge of being crowded out of the picture altogether. This cannot be allowed to happen.

To make space for the historical imagination to flourish, it will be necessary to develop an appropriate visual and written language for the exercise and development of the historical or geometrical imagination. Perhaps a reconstructed verbal language such as Lefebvre calls for would be adequate to the task. To construct such knowledge, it is necessary to examine and perhaps even discard many apparently useful but ultimately limiting notions, including, for example, a "quasi mechanics of history." Although the metaphor of "mechanics" has so far supplied a convenient point of departure, it is important to recognize that history (like electronics[10]) may not be explicable on the basis of Ortega y Gasset's mechanical metaphor. If history has nothing to do with mathematics, it may have no more to do with mechanics. If history is indeed the study of problems of space, however, the next logical step is to investigate the history of problems of space through a metaphor that will lead to deeper understanding of how we identify, communicate, and solve problems of space jointly through language, mathematics, and graphics. These knowledges actually work together in producing and inscribing meaning through *parable*, and the relation between words, numbers, and pictures will be seen to be fundamentally parabolic in character. Similarly, the workings of the historical and geometrical imagination are also parabolic in character. Examining the nature and importance of parable, and the parabolic construction of both knowledge and social space, is the next step toward gaining a better understanding of the role of thinking geometrically in producing space.

CHAPTER 3

STORIES OF SPACE

The role of geometrical thinking in imagining and producing social space is demonstrated and documented by inscription—words, numbers, and pictures—which serve as *parables*, or stories, of space. Conventionally, a parable is a "short, simple story from which a moral lesson may be drawn . . . usually an allegory" (*Websters*, 1983:1297). To account for the power of parable this definition needs itself to be drawn out in two respects. First, the word "drawn" suggests more than "extract"; it serves also as a metaphor for visualizing a relationship. The role of imagery—though in a more figurative than literal sense—is crucial in parable. Even though almost nobody except those working in the practically obsolete genre of allegorical painting make actual pictures of parables, parable involves picturing, at least in the mind's eye. Even though parables are constructed with words, they get at something that resists being reduced to words. Second, the "lesson" of a parable may not always be a "simple" or even a "moral" one. The work that parable performs is far more complex and significant; it is actually a demonstration of what I call thinking geometrically.

Mark Turner (1996) defines parable as a "small spatial story" (163) based in "image schemas": "skeletal patterns that recur in our sensory and motor experience" (16). These schemas, or patterns, serve as the basis for recognizing and categorizing objects, events, and sequences in the world. Parable is "the laboratory where great things are condensed in a small space" (5). It thus involves a particular or "projective" relationship between objects, events, and sequences and the "story."

Etymologically, "projection" is from the Latin *projectus*, "to throw forward" (*Websters*, 1983:1439). This sense is preserved in predicting the future of, for example, weather and economics, where the word "projection" usually means forecasting: literally, a "throwing forward"; figuratively, a conjecture as to future events or conditions—that is, a possible future path extrapolated from a knowledge of the present. Unlike the trajectory of a moving object such as a cannonball (bound by the laws of physics), which (if relevant environmental conditions such as wind are accounted for) can be determined exactly from the moment of its release, a weather or economic forecast, like any cultural forecast, is never entirely precise or accurate. Weather or economic trajectories, as expressed in graphs, are only entirely accurate when they are made retrospectively, and, as every mutual fund prospectus acknowledges, past performance is no guarantee of future results. Projection is thus, at least in part, conjecture about the future of relationships.

But more significantly, projection is the capacity to "map" one kind of structure (or image schema) onto something entirely different, thereby aligning them both in a certain way. To put it another way, projection draws a connection between two different structures. Projection often works by carrying a structure from a source we "understand," that is, an already constructed projective relationship, to a target we want to understand. Thus, projection involves relationships between objects, events, and sequences in the world as well as between other stories or parables.

The process of parable involves increasingly complex processes of projection. A word, name, sentence, or story projects an object, event, or concept. That projection also projects onto other words, names, sentences, or stories in an increasingly complex network of projective relations. A word or name is a small parable. But even in the case of words or names, with their potential for multiple mappings, the projective relations can be rather difficult to untangle. For example, even the word "parable" projects differently in various traditions and cultures; this is of course the challenge of translation. Interestingly, the Spanish word *parábola* means both story and geometrical curve. A sentence is a somewhat more complex parable. A sentence, as a series of words strung together like "Mary threw the rock," might also be considered a "line" of thought, which might be straight, circular, or curved. Curved lines of thought are quite common. Longer stories or tales, such as the fables that Shahrazad tells in *The Thousand and One Nights* and examined by Turner, are much more involved parables with many more dimensions of meaning. These multiple dimensions of meaning construct multiple and intersecting lines of thought that are contextualized parabolically[1] through both other parables and experience in the world. The greater the projective multiplicity, the more difficult the parable can be to untangle. Thus, parables are subject to endless reinterpretation.

The process of parable always involves selection, reduction, and a positive process of making. A parable extracts; it is never completely inclusive of whatever it projects. A story never tells everything that happened; it selects and focuses on whatever is most meaningful or important. Neither is a parable, strictly speaking, representational. Rather, it is an implicit statement of difference, because events and objects in nature are not reducible to the language we use to identify, categorize, and relate to them. In so selecting and reducing, parables make meaning, or draw alignment, rather than being a passive vehicle for it. In short, they project.

Making connections among image schemas is the basic "move" of parable that makes language possible. Turner's account of the development of language inverts the notion that parable is a product of narrative. According to Turner, it works the other way around: "The dynamic processes of parable are basic to the construction of meaning and the construction of language. Story precedes grammar. Projection precedes grammar. Parable precedes grammar. Language follows from these mental capacities as a consequence; it is their complex product" (168).

Despite the centrality of visual and spatial concepts—such as image schema and the fruitfully ambiguous extension of parable into parabolic—in this account of language, Turner asserts that the mind is fundamentally literary in character. I do not agree. Even though language may be a defining characteristic of what it means to be human, not all cultures have a literature, that is, unless one is prepared to accept the oxymoronic label "oral literature."[2] Writing, a prerequisite for literary thought, is actually a recent development in human experience, and literacy is as yet by no means universal. The mind, however, is fundamentally parabolic.

Writing, like grammar, is a complex product of parable. Even though the letters of the English alphabet bear a parabolic relation to the sounds that we make when speaking, this relation is almost entirely abstract. Even though letters, as graphic entities, are obviously visible, they do not usually look like the sounds they represent. And except in the case of onomatopoeia, words don't sound like their subject matter. Writing, that is, literature, transforms the ways we use language and makes possible systematic forms of thought, like logic and philosophy (Ong, 1967:1982). Even so, however fundamental the literary mind and its products may be to contemporary Western technological culture, literacy is neither a universal nor a necessary requirement for civilization in general.[3] There are many other highly developed forms of thought that may be parabolic in character but are distinctly nonliterary, such as mathematics.

Even though Turner's focus is directed at the literary mind, the notion of projection opens up a connection (projects) to other forms of thought and expression. Words, numbers, pictures, and the stories that are composed from them constitute new objects and events in parabolic relations both to their subject matter and to one another. They project across different levels, in many different directions, producing a space of meaning. In this multidimensional space, parables themselves become subject matter for parable. Parables, as objects in this space, are multidimensional and spatial in character. As such, they have much wider meaning than can be understood through spoken or written language alone. Parable is not bound exclusively to language, either spoken or written. Parable, as projection, is an essentially spatial phenomenon and might be better understood and expressed through other means.

Both language and mathematics are among the many possible extensions of parable. But despite some equivalences between them, language and mathematics each constitute a different parabolic space. For example, we have convenient names for most basic mathematical entities and procedures. Integers can be readily understood as simple parables: "One" stands for a unit, "two" stands for a couple, and so on. To "add" means to accumulate, and to "subtract" means to diminish. One might conceive of an "equation" as the mathematical equivalent of a sentence; both are small spatial stories. However, even though we can describe such operations in language, the rules of language—such as grammar and syntax—do not really apply to mathematical procedures. Performing mathematical operations such as counting words in a sentence tells us nothing about its meaning.

Mathematics projects parabolic structures and relations that are different from those projected by language. For example, cut an apple down the middle. You have two half-apples. You also have two new units, with a definite relation to the original unit. Fractions (ratios) such as 1/2 not only mean "one-half" of something, but also, and more importantly, assert the existence of mathematical relationships that do not depend on their original relation to material objects or to their expression through language. A theoretical basis for advanced mathematics (beyond elementary arithmetic) emerges from extending this fundamental parabolic relation of number to object and recognizing that numbers are themselves objects. Even though mathematical entities, relations, and operations are more often than not "explained" through resort to language, proficiency in mathematics—"numeracy"—requires moving beyond this elementary knowledge and internalizing mathematical principles and procedures that cannot be reduced to words. Mathematical operations often have no exact parabolic equivalent in language. For example, in mathematics, "multiply" does not involve the generation of new objects, as it does in the world of human affairs, in, say, reproducing the species. Rather, it is a specific statement of relation that implies its opposite—"division"—in a very precise sense.

Without the development of a powerful and flexible notational system—which includes symbols such as + & # *—/ =, Arabic numerals and the zero—that is distinctly different from ordinary language, modern mathematics and the science and technology based on it would not be possible. Through such relations, mathematics opens up a whole new field of meaning that is ultimately inaccessible through language alone and can only be understood by the study and exercise of mathematics. Numeracy opens up a whole new kind of awareness, which cannot be fully explained in language, even though language is involved to a certain extent in the development of mathematical knowledge. Although mathematical expressions of a more complex (algebraic) kind are bound up with language through the alphabet, the letters of the alphabet no longer stand for sounds but for unknown quantities or variables ($x + 5$ is a simple example). Through equations, which might be construed as a mathematical kind of sentence, the parabolic equivalence and interchangeability of numbers and alphanumerical expressions are made explicit and precise in ways that language does not allow. In poetry, for example, the arrangement of words is somewhat flexible, but words are not completely interchangeable in the production of meaning. However, in mathematics, at least according to certain rules, symbols can be rearranged with no change in meaning (for example, $x + 5$ is equivalent to $5 + x$, and $x + y = 5$ is equivalent to $x = 5 - y$).

All of mathematics would be no more than an elegant intellectual parlor game were it not for the fact that through mathematics we can consistently and accurately measure and describe the behavior of objects in the world in ways that are impossible through other means. Even though mathematics may have emerged from an empirical basis (the need to count things, perhaps), pure mathematics ultimately admits to no necessary connection to objects, events, or causes in the real world (Kline, 1985). However, the making of such connections—parables—is of fundamental importance to human affairs, and the fact that such connections can be made is one of the reasons that mathematics, science, and technology have come to dominate our lives. For example, the surveying of land in ancient Egypt may have been based on the empirical knowledge that a length of rope divided into lengths of 3, 4, and 5 units will enclose a right triangle, but it took the Pythagorean discovery and proof of the relation $x^2 + y^2 = z^2$ to elevate this knowledge into theory and generalize its application to other and more complex problems of pure theory in mathematics or science: for example, Galileo's formula for describing the behavior of falling bodies ($d = 16t^2$, where *d* is distance and *t* is time) or Einstein's more famous equation for the relation of matter to energy ($E = mc^2$). Both of these are actually precise and systematic forms of parable.

Language and mathematics are such well-developed systems of meaning that it is easy to overlook how literacy and numeracy actually depend on the embodiment of words and numbers in graphic form. As already noted, letters and numbers are not really pictures of what they stand for even though they embody and demonstrate projective relations. The origin of projection in image schemas suggests that projection might be best understood through the study of how we parabolically construct and use images. However, image schemas have no physical existence. Pictures, as representations of image schemas, are much more convenient.

A simple picture, such as a line drawing, is a parable. Simple drawings project recognizable shapes such as outlines to represent real objects. More complex drawings project spatial relations among multiple objects and sometimes even multiple points of view. Drawings are conventionally useful because they show what things "look like" (especially modern perspective drawings). But more important, as Mitchell (1994) points out, pictures not only represent and comment on objects and events in the world, they represent and comment on each other. "Metapictures" establish intelligible parabolic relations with other pictures. Metapictures also comment on the parabolic relations they establish between themselves and other modes of communication such as language. More often than not, pictures demand not only visual attention but a verbal response. It is difficult not to comment (favorably or unfavorably) on what one sees when viewing a painting, even though this may not lead to any conclusive understanding of the parabolic relation between words and pictures, as hundreds of years of art criticism demonstrates.

Most important, pictures make projection explicit in ways that words or numbers cannot. However, reading graphic evidence is a much more difficult and inexact process than reading textual or mathematical evidence. It requires disciplined, but nonetheless parabolic, imagination and (re)construction. This is illustrated by the example of assembling an understanding of an object from the partial and limited views seen in an engineering drawing. What makes reading such a drawing possible at all is a formal knowledge of the system of projection used in its making: specifically, orthographic projection, which will be discussed in chapter 4. What makes the development of such a system possible in the first place is parable.

In combination with mathematics and language, the graphic concepts of point and line (which are imaginary entities, having no existence in natural space) emerge and serve as the basis for more complex relations such as intersection, parallel, and angle. Even though these parabolic relations project onto other modes of communication (language, in particular, is shot through with graphic and spatial metaphor, especially that based in geometrical perspective), the meaning of pictures cannot be contained in, or explained by, either language or mathematics.

Even though the language of mathematics has served to transform our understanding of physical space, it has yet to tell us much about, or even inform our understanding of, social space—the nonmathematical space(s) of everyday life. Both mathematics and spoken and written language serve poorly to identify and describe how we produce this space on a daily basis through the patterns of our lives. The mundane behavior of our own minds and bodies is still as much a mystery to us as the arcane behavior of celestial bodies was to the ancients.

The success of mathematics as the language of science in explaining the mysteries of the natural world leads one to wonder about the possibility of developing a similarly precise language—such as Ortega y Gasset's quasi mechanics of history—with which to describe the parabolic trajectories of human bodies and events as accurately as mathematics describes the trajectories of celestial bodies and events. Graphics may be more useful than either language or mathematics in developing a systematic, precise, and explicit understanding of the central importance of parable and projection in the construction of knowledge.

Parable is what connects all dimensions of knowledge. Knowledge is constructed through parabolic and projective relations. Language, mathematics, and graphics do not constitute distinct spheres of knowledge and experience, even though they do constitute different frames of reference. Instead, language, mathematics, and graphics are each no more than limited and partial (two-dimensional) representations of a larger and more comprehensive space that is never totally reducible to any representation in a simpler dimensionality. Modeling this space is an exercise in geometrical thinking.

History is also a parabolic construction. It's one thing to recognize the potential of geometrical thinking in the construction and history of knowledge, and another thing to develop this potential. If we are to "make space" for the historical imagination, we can't really wait for a Galileo of the historical sciences to arrive on the scene; what we really need to do is make the exercise of the historical, parabolic, and geometrical imagination a widespread and commonplace practice.

The act of geometrical thinking in a responsible manner carries with it a recognition of how we are continuously and actively engaged in parabolically constructing and modeling ourselves and the world around us. We both produce, and are produced by, our involvement in the world. At the same time that we are so involved, it is also necessary to be able to somehow get outside the constraints of this involvement and look at things from a point of view of a more complex dimensionality. How this might be done can be gained from a study of two extended parables: Edwin A. Abbott's short novel *Flatland: A Romance in Many Dimensions* and Plato's allegory of the Cave. Although the connections I wish to draw between these works are my own, I am not the first to make them. The mathematician Rudy Rucker, in *Geometry, Relativity and the Fourth Dimension* (1977:4), remarked that Plato's allegory "can be seen as prefiguring the concept of Flatland," but he did not go on to explore this connection. The projective relation between them deserves further investigation and is essential to my argument that developing the ability for thinking geometrically is crucial in understanding our own parabolic imaginations as well as understanding how we both produce, and are produced by, (social) space.

The Story of Flatland

At the turn of the twentieth century, conjecture as to the existence of a fourth dimension of space fueled considerable philosophical, spiritual, artistic, and mathematical inquiry. Art historian Linda Dalrymple Henderson (in the introduction to Robbin, 1992) describes how a concern with the fourth dimension was a central theme in early modernism, but after Einstein's theory of relativity posited time as the fourth dimension, interest in a fourth spatial dimension waned. However, recent work, free of much of the spiritualistic and mystical fervor that motivated earlier speculation about the fourth dimension, resurrects the notion that we need to learn how to think in more than the three dimensions that are the basis of everyday thought and action. Today, with the powerful new visualizing tools we have in computer graphics, artists and scientists are renewing the investigation of problems of physical or mathematical space in four or more dimensions.[4] Mental space, the space of the imagination, is now recognized as having more than the three dimensions of conventional Cartesian space. For example, psychologist Howard Gardner's (1983/1993; 1993) theory of "multiple intelligences" posits at least seven different yet interpenetrating dimensions of intelligence (see Chapter 2, footnote 7).

Thinking geometrically as I conceive it is a combination of at least three of Gardner's dimensions: logicomathematical rigor, spatial imagination, and kinesthetic awareness. Although linguistic skill is a powerful component of all human action, I believe that these three dimensions can exist independently of, and perhaps even prior to, verbal thought and action. Further, thinking geometrically might best be conceived as a metaintelligence, in that it is not a singular, or even dominant, intelligence in the sense that linguistic or logicomathematical intelligence is. Thinking geometrically is instead the ability to draw on and coordinate several dimensions at once.

The inscription alleged to have been engraved over the portal of Plato's academy—"Let no one enter here who is ignorant of geometry"—is firm testimony to the early Greeks' recognition of the importance of geometry over other forms of intellectual (philosophic) endeavor. However, their promising discoveries were to be little improved on for over two millennia, not until non-Euclidean geometries emerged in the nineteenth century. The impact of these radical new geometries on (social) conceptions of space sparked the popular imagination enough that Abbott's initially pseudonymous story, *Flatland* (1952), originally published in 1884, was an immediate success. It has remained in print for more than a hundred years. His story is well worth reading for its many merits; here, I can incorporate only a limited portion of its subtle wit and still-relevant cultural commentary. Most important for the present undertaking, Abbott's story tells us a great deal about the limitations of language in dealing with the challenges of geometrical thinking as well as the how these limitations persist over time.

The "Author"[5] of Abbott's unnamed fictional first-person narrator is an inhabitant of Flatland, which is a planar world, inhabited by two-dimensional beings called, not surprisingly, Flatlanders. The narrative is a transcription of the Author's memoirs, written in prison. Why he is there will be discussed later. The Author describes the physical features and social hierarchy of the population of Flatland in familiar English (and Euclidean) terminology:

> Our Women are Straight Lines.
>
> Our Soldiers and Lowest Classes of Workmen are Triangles with two equal sides, each about eleven inches long, and a base or third side so short (often not exceeding half an inch) that they form at their vertices a very sharp and formidable angle. Indeed when their bases are of the most degraded type . . . they can hardly be distinguished from Straight Lines or Women; so extremely pointed are their vertices. . . . These Triangles are distinguished from others by being called Isosceles. . . .
>
> Our Middle Class consists of Equilateral or Equal-Sided Triangles.
>
> Our Professional Men and Gentlemen are Squares (to which class I myself belong) and Five-Sided Polygons or Pentagons.
>
> Next above these come the Nobility, of whom there are several degrees, beginning at Six-Sided Figures, or Hexagons, and from thence rising in the number of their sides until they receive the honourable title of Polygonal, or many-sided. Finally when the number of the sides becomes so numerous, and the sides themselves so small, that the figure cannot be distinguished from a Circle, he is included in the Circular or Priestly order; and this is the highest class of all. (9)

We who live in the three dimensions of what Abbott calls Spaceland can perhaps most easily visualize such beings as shadows, but how they would see themselves, or us, is quite a different matter. To put ourselves in their shoes, so to speak, requires a considerable stretch of the imagination, though perhaps not as much as would be required of them to imagine our world. Trying to do so is useful, however: first, as an exercise in visual imagination, and second, as a step toward developing a better understanding of the real interdependencies and limitations of language and visual representation that are demonstrated by the Author's experiences.

Flatlanders, being restricted to two dimensions, cannot perceive the universe according to our standards of experience. The two dimensions of Flatland permit its inhabitants no more than edge views of each other. Seeing the actual outline of a planar figure is only possible from a point of view outside its plane, which is of course a physical impossibility for Flatlanders.[6] An edge view of any linear, polygonal or circular shape is, of course, a straight line, the exception being an end view of a line, which is a single point. Flatlanders' different shapes can only be inferred by one or several of three methods: by *feeling*, that is, touching each other's vertices, very carefully, so as to avoid being inadvertently impaled by a sharp angle; *hearing*, as each class has a characteristic manner of speaking, so that the socially awkward act of feeling is not usually necessary; and most difficult of all, *seeing*, that is, observing the rate at which the more distant sides and vertices of a shape fade away from vision, in the same way that more distant objects in our space appear to be less distinct than those close up. As a mode of perception, seeing is actually assisted by the chronic existence of a Fog, which enhances their version of what we call atmospheric perspective. The thicker the Fog, the further the distance of what is perceived. The difficulty of seeing makes it a specialized and elite practice, restricted to the higher classes.

The Author's first inkling of other worlds comes in the form of a dream vision of Lineland, which has only one dimension. Lineland is even more dimensionally restricted than Flatland. Each of the inhabitants is either a point, or a line of varying length, and the only possible motion they enjoy is back and forth along the single dimension that constitutes their world. Seeing is even more limited in one dimension than it is in two, and sound is the Linelanders' only reliable means of communication.

In his dream, the Author enters this space and determines that the longest Line is actually the Monarch, or King, of Lineland. The Author's sudden and mysterious presence throws Lineland into turmoil. In an effort to explain himself and his origins, he attempts to explain the existence of a second dimension to the King by stating that another direction of motion exists: specifically, left and right movement, which is perpendicular to the single axis of linear movement that characterizes Lineland. The King's response to this revelation is both foreseeable and prophetic of the Author's later experiences:

> King: Exhibit to me, if you please, this motion from left to right.
> I: Nay, that I cannot do, unless you could step out of your Line altogether.
> King: Out of my Line? Do you mean out of the world? Out of Space?
> I: Well, yes. Out of *your* World. Out of *your* Space. For your Space is not the true Space. True Space is a Plane; but your Space is only a Line.
> King: If you cannot indicate this motion from left to right by yourself moving in it, then I beg you to describe it to me in words.
> I: If you cannot tell your right side from your left, I fear that no words of mine

can make my meaning clear to you. But surely you cannot be ignorant of so simple a distinction. (62)

Even though speech, being linear, is as primary a means of communication in Lineland as in Flatland, words fail both the Author and the Monarch. The dream becomes chaotic, and the frustrated Author is forced to leave Lineland for his own safety. He awakens, with relief, to the breakfast bell in Flatland. But his dream turns out to be a sort of premonition. Later that day, while indulging in his hobby of studying geometry, the Author is visited in turn by a three-dimensional Stranger, who proves to be as challenging and confusing to the Author as he himself was earlier to the Monarch of Lineland. After awkwardly "feeling" (in the sense of touching) this Stranger, who turns out to be the most perfect Circle, the Author cannot comprehend that his visitor is actually a Sphere, from a Space of more dimensions than two:

Stranger: Pooh! what do you know of Space? Define Space.
I: Space, my Lord, is height and breadth indefinitely prolonged.
Stranger: Exactly: you see you do not even know what Space is. You think it is of Two Dimensions only; but I have come to you to announce to you a Third—height, breadth, and length. . . .
I: Would your Lordship indicate or explain to me in what direction is the Third Dimension, unknown to me?
Stranger: I came from it. It is up above and down below. (69–70)

After prolonged difficulty in explaining the third dimension and his nature as a solid body, the Stranger resorts to demonstration, moving in and out of Flatland much as the Author tried to demonstrate his own two-dimensionality to the Monarch of Lineland. The Stranger then appears to be a circle of varying diameter, as demonstrated in figure 3.1.

This phenomenon deserves some explanation. Three-dimensional objects passing through a two-dimensional plane generate sections, which are technically different from shadows, which are projections. It is doubtful that Flatlanders could appreciate the difference, because both would appear as lines in edge view, the only view possible in their planar space. Sections, unlike shadows, are not usually part of our ordinary three dimensional experience: their status is more figurative, or theoretical, than literal. In common physical experience, we have all seen cross sections, for example, in the form of slices through trees. But this perception is of a slightly different order than a graphic construction of a section, which is a partial and abstract representation of the insides of a solid object. The physical integrity of any object is destroyed as it is sectioned, but to an engineer or designer, a section is not an actual cut. Rather, "cutting a section" on paper is a way of seeing inside things, of representing internal relationships that could not be observed physically without actually cutting through the object itself. Magnetic Resonance Imaging (MRI) is just such a technique of sectioning the body through noninvasive computer technology, which is considerably less expensive and less traumatic than the physical alternative of exploratory surgery.

If a human body were to pass through Flatland—a phenomenon similar to being gradually lowered into water, the still surface of which represents a two-dimensional plane—the result would be a changing series of sections. Starting with the feet, the first sections would be 12 irregular closed curves, the outlines of 10 toes and two soles. These sections would quickly become (or morph into) two separate closed curves at the ankles. These would become one as the sectioning plane reached the hips. Before this, however, assuming that the arms were hanging at the side, additional sections would suddenly appear. If the fingers were apart, as many as 10 more sections could appear! These curves would in turn become one as the section plane reached the shoulders. At the head, assuming one wasn't completely bald, this section line would appear to be surrounded by a cloud of points: hair has practically no width, and a section through a line is a single point. Could a Flatlander even begin to comprehend that these changing sections and points were all related expressions of a physical unity, a single, three-dimensional body?

The difficulty of unifying such perceptual differences into an inference of an object of other dimensionality can perhaps be more clearly understood by examining the behavior of simple two-dimensional shapes, such as circles, in spaces of varying dimensionality. The line produced by a circle passing through Lineland could only be seen in end view as a single point. When a circle lies entirely within the planar space of Flatland, it can be seen only in edge view, as a curved line, which can readily be distinguished from a straight line, especially with the assistance of the chronic Fog of that space. However, a circle that passes at an angle through the plane of Flatland could be seen only as a straight line or, at the least, a single point of tangency if it just touches the plane, which is identical to the results of a circle that passes through the single dimension of Lineland. In any case, a complete perception of circularity can only be theorized, and not experienced, in spaces of fewer than three dimensions.

When passing through a line, a solid sphere, like Abbott's Stranger, will appear first as a point that immediately grows into a line of varying distance. However, when passing through a plane, a sphere starts as a point and then appears to be a circle that grows until its equator is reached, after which it shrinks until it turns into another point and disappears altogether, as figure 3.1 illustrates.

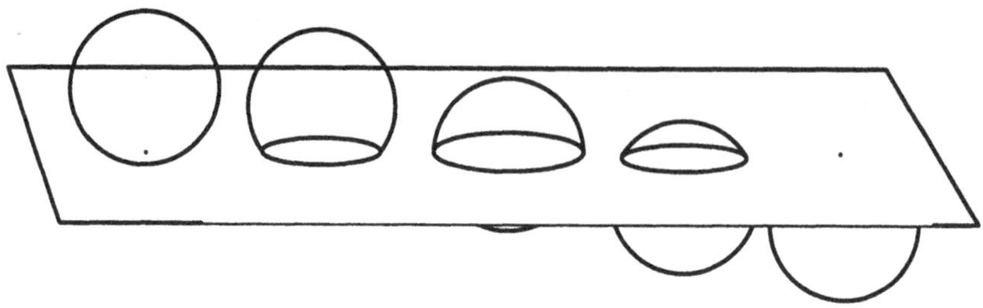

Figure 3.1

In our three dimensions, linearity, circularity, and sphericity are directly observable, even though they may be distorted. For example, seen at an oblique angle, a circle appears to be oval (as in the oblique plane shown in figure 3.1), but we automatically accommodate this in our perspectival way of seeing the world. Linelanders or Flatlanders, lacking a point of view outside their own plane, must see such shapes passing through their own more limited spaces as unpredictably changing linear phenomena—sets of points, or lines of varying length and shape—rather than outlines, which can only be inferred by Flatlanders through the same kind of laborious intellectual effort that we must pursue in trying to comprehend things that exist in four or more dimensions.

Thus, it's no surprise that the Stranger's demonstration confounds the Author. As a last resort to communicate what he calls "the Gospel of the Three Dimensions," the Stranger seizes the Author and carries him out of Flatland into the third dimension. After overcoming his initial disorientation, the Author begins to see in three dimensions:

> It was even as the Sphere had said. The further we receded from the object we beheld, the larger became the field of vision. My native city, with the interior of every house and every creature therein, lay open to my view in miniature. We mounted higher, and lo, the secrets of the earth, the depths of mines and inmost caverns of the hills, were bared before me.
>
> Awestruck at the sight of the mysteries of the earth, thus unveiled before my unworthy eye, I said to my Companion, "Behold, I am become as a God. For the wise men in our country say that to see all things, or as they express it, *omnividence*, is the attribute of God alone." (82)

Abbott's narrator, through the fantastic intervention of a three-dimensional being, manages to attain a point of view outside his own space of Flatland and thus learns to think and see in more dimensions than his fellow Flatlanders can even begin to conceive.[7] His new-found ability for thinking geometrically, gained through his direct experience of a third dimension, leads him to theorize further and comprehend the possibility of a fourth dimension. He is intoxicated with his discoveries.

Despite his exuberance, the poor Author is chagrined to find that this powerful new knowledge is poorly received by his fellow Flatlanders. Much as he lacked the means to communicate right and left to the Monarch of Lineland in his dream, he lacks the verbal or visual means to communicate the up and down of the third dimension to Flatlanders. Now bound by the spatial constraints of his own two dimensions, he cannot loft them to the heights he experienced by traveling with the spherical Stranger. In fact, his knowledge is a threat to the carefully developed order of Flatland. For evangelizing the Gospel of the Three Dimensions, he is imprisoned for life, but spared from the death penalty because of the lunacy of his claims. The final words of the Author's account eloquently summarize his pathetic plight:

> Hence I am absolutely destitute of converts, and, for aught that I can see, the millennial Revelation has been made to me for nothing. Prometheus up in Spaceland was bound for bringing fire down for mortals, but I—poor Flatland Prometheus—lie here in prison for bringing down nothing to my countrymen. Yet I exist in the hope that these memoirs, in some manner, I know not how, may find their way to the minds of humanity in Some Dimension, and may stir up a race of rebels who shall refuse to be confined to limited Dimensionality. (102)

At the risk of being labeled one of a "race of rebels" (or "off the wire" like Jason Bourne), it seems to me worthwhile to resume the Author's quest for the Golden Fleece of seeing from more complex levels of dimensionality, tempered by the advance knowledge that there is no one view sufficient to take in a whole world. Neither Lineland, Flatland, Spaceland, nor any other world of other dimensionality can be seen or depicted from a single point of view within that space. Omnividence is an impossibility, a perspectival illusion. Even the Author's overview of Lineland, impossible to comprehend for Linelanders, would have had to fade at the extremes, like the chronic Fog that aids near vision in Flatland but obscures objects more distant. The gradual loss of detail of distant objects in any dimension precludes their being completely represented in any graphic or verbal form.

We in Spaceland already live out the Gospel of Three Dimensions in daily existence. For us, the challenge is to expand our vision to four or more dimensions, beyond the constraints of our existing knowledge and bodily experience. If we can, through a little effort, imagine Flatlanders in a world of one less dimension than ours, we ought also be able to conceive of inhabitants of a world of one more dimension. This four-dimensional world, which has been termed Hyperspace, would have to be at least as incomprehensible to us as our three-dimensional world would be to Flatlanders. For example, what are we to make of the possibility of four-dimensional, hyperspatial objects projected or sectioned into our own Spaceland? Or, to stretch things a little further, what are we to make of the possibility of an infinite number of other (parallel) three-dimensional spaces, worlds unavailable to our perception except through sections and projections, but which are readily visible and comprehensible from the point(s) of view of a Hyperspacer? (Hyperlander sounds too three-dimensional!)

We don't have the perceptual ability to see hyperspatial objects in four dimensions, any more than Flatlanders have the ability to see in our three. However, I like to think that we have more power and flexibility than Flatlanders in understanding an extra dimension or two, since we can see Flatland (even though it is often difficult) from outside their world, but they can only theorize about us. To theorize a fourth dimension, however, we must move beyond an exclusive reliance on language as a tool for the spatial imagination.

Speech is a primary means of recognition and communication in Flatland, just as it is for us. But what would writing mean to a Flatlander? Script, or print, as we know it, would be unintelligible to Flatlanders, since they can't view it perpendicularly to the page in three-space as we do. At most, in Flatland, one could write language as a series of points and lines; either Morse code or modern bar codes could serve if viewed at the right angle, that is, perpendicular to the line of code. If Flatlanders could ever develop a graphics, that is, some system of visual representation, it would have to use the same zero- or one-dimensional elements as text (points and lines) which would be hardly distinguishable from the points and lines that represent real objects. This limitation is crucial. Indeed, one of the reasons Abbott's Author was imprisoned was his inability to provide either visual or verbal evidence of his claims for the existence of a third dimension.

Perhaps the universal human experience of language—generally, one-dimensional (in oral experience), and at most, two-dimensional (in text)—similarly limits our perception of both the two- and three-dimensional worlds in which we live, limiting as well our imagination of spatial dimensions other than the three of conventional physical experience. "Plain speaking" here becomes more than just a metaphor or a bad pun! From the point of view of one locked into the plane of language, that which is outside language is outside of the horizon of experience. Even exploring other languages doesn't open up this horizon so much as it projects the horizon onto alternate planes. These planes are equally and materially inaccessible except as limited projections or translations onto the plane of the other.

Abbott's work is more than a thought experiment or a mathematical whimsy, in the same way that Lewis Carroll's *Alice's Adventures in Wonderland* is more than a child's fantasy tale. It is a parabolic adventure in logic and illogic, language, and geometry. Although Abbott tries to discourage such a reading in his preface, it seems reasonable to read *Flatland* as a commentary on, or even a satire of (like Swift's *Gulliver's Travels*), our habits of perception and behavior, which are clearly demonstrated in the patterns through which we speak, read, and write. In adopting a response to Abbot's commentary or satire, however, it is important to acknowledge that language is, even in our own dimensional space, an unwieldy instrument by which to explore problems of geometry and spatial perception. Communicative (spatial) practices such as speech, writing, and graphics, as they stand, are inadequate for the project of remodeling themselves. Speech, whether in person or via electronic media, in our world as in Flatland, is essentially linear. All other mass media such as print and television are at best two-dimensional spatial phenomena. In various incarnations, such as the TV screen, pages of a book, posters and flyers, maps, photolithographed paintings, and so on, mass media limit what we know through three-dimensional experience to a series of two-dimensional representations. No one of them is complete, just as no one map is sufficient to describe a whole three-dimensional object such as the globe. Distortions and omissions are inevitable (Wood, 1992). Even topographical maps, which express the altitude of features on the landscape, tell us little about what that landscape really looks like from our horizontal, perspectival point of view. Even though we become accustomed to mentally flipping the bird's-eye point of view of the map to the way it is seen from the ground, this is an often difficult and error-prone process: spatial reversals and disorientation result. Engineering drawings, which provide several orthogonally related points of view as a way around this difficulty, present in turn the difficulty of integrating these multiple two-dimensional views into an imaginary conception of an object in three dimensions and can be similarly disorienting, especially in regard to complex structures. Either way, we are faced with the intimidating prospect of somehow translating, transforming, or integrating partial (two-dimensional) graphic representations of real or imaginary objects to gain an understanding of the three (or more) dimensions of an object or relation. By insisting on language, a two-dimensional medium (at best) as the common denominator of communication, we are all Flatlanders at heart.

Generalizing from two dimensions into three suggests a way toward stepping outside our own three-dimensional spatial limitations to deal with four or more dimensions of space. For example, what if we were to reconceive three-dimensional objects, which we conventionally think of as wholes, instead, as sections, or projections—partial representations—of objects that may exist completely only in four or more dimensions (a notion suggested by Rucker)? The previous examination of Linelanders' and Flatlanders' experience of objects of greater dimensionality—things that exist outside their own spaces and appear to vary as they pass through a space of lower dimensionality—suggests that perhaps objects or relations in our own space that appear to vary from different points of view (or at different times) and are differently named might be recognized simply as limited two- or three-dimensional sections or projections, different parabolic manifestations of a single object that exists in a space of other dimensionality. To explore this prospect requires an expansion of language and representation beyond their current limits as partial and at most two-dimensional systems of communication. It also requires the definition of key terms such as "dimension."

In Euclidean space, a dimension is defined mathematically as a perpendicular direction from some other dimension. This can be demonstrated graphically (after Rucker [1977], whose exposition is based on Bragdon's [1913] work). A point dragged in an unvarying direction along some distance through space defines a straight line, which defines one dimension. Because the graphic space of the page does not permit representing the infinite length of a dimension, here I will represent it as a line of arbitrary length (figure 3.2).

If a line is dragged through space in an unvarying direction (other than in the direction of travel), the result is a flat plane. If this direction is perpendicular to the line, and the distance moved is equal to the length of the line, a square can be created by connecting the ends of the lines, as shown below in oblique view (figure 3.3).

If a planar shape such as a square is dragged through space in an unvarying direction perpendicular to the plane, the result is a solid, in this case, a cube (figure 3.4).

Any of the corners of a cube could serve to describe the three perpendicular axes or dimensions of three-dimensional space (conventionally labeled X, Y, and Z). These coordinate systems would differ only in regard to the location of the origin, or intersection, of the three axes and their directionality, that is, whether an object is situated in positive or negative coordinates. In conventional

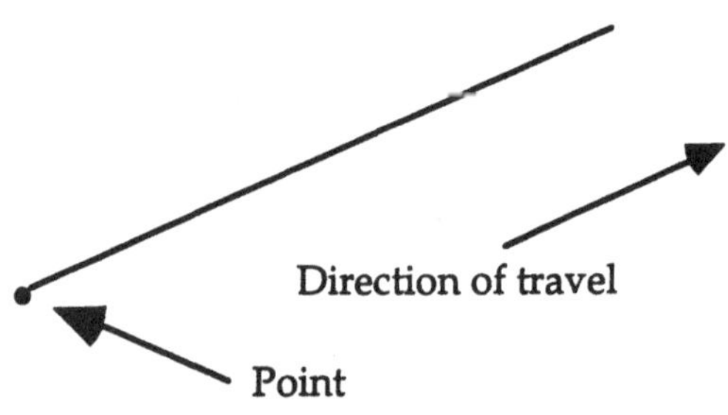

Figure 3.2

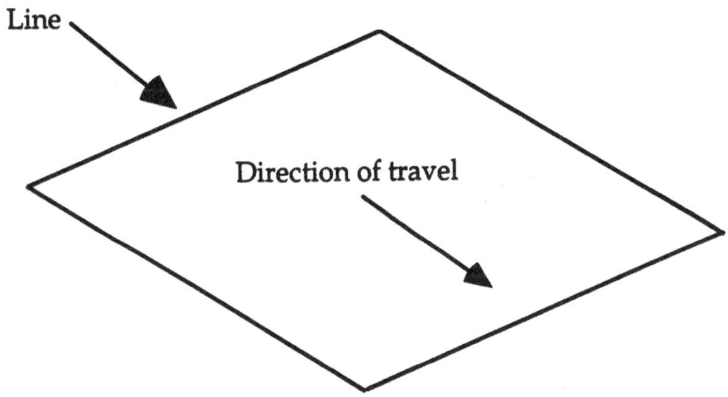

Figure 3.3

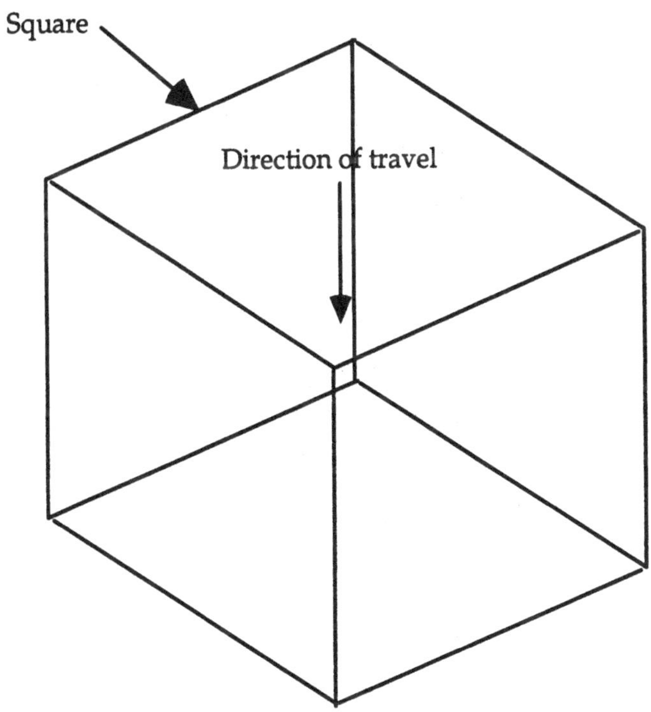

Figure 3.4

Cartesian graphic representation, positive X is to the right, positive Y is up, and positive Z is the "missing axis," which points toward the viewer, away from the picture plane. In ordinary human experience, these dimensions are related to the axes of the body: right-left, up-down and forward-behind. It is interesting that right-left and up-down coincide with the landscape orientations of horizontal and vertical, but I know of no such analogous term for depth, or the Z axis, which suggests that our experience of three dimensions is routinely reduced to a two-dimensional conceptual field.

The three axes of Cartesian space seem to exhaust the possibilities of further spatial subdivision. In what direction would a fourth perpendicular go? One must proceed on the basis of imagination and theory. It is a direction perpendicular to three-space. Even if it can't be imagined, it can be demonstrated graphically. Just as a line was developed by dragging a point in one dimension, a square was developed by dragging a line in two dimensions, and a cube was generated by dragging a square in a third dimension, one could similarly drag a cube in a fourth perpendicular direction and then connect all the resulting corners (figure 3.5). The result is a four-dimensional solid, usually called a "hypercube." Just as a square is shown by four perpendicular lines, and a cube is described by six perpendicular squares, a hypercube is described by eight ordinary cubes (one new cube extruded from each face, plus the original cube and its duplicate), each perpendicular to one another. This can be shown in two dimensions through the graphic device of foreshortened angles. For an interesting mental exercise, try counting the cubes!

The process of extrusion can be continued indefinitely into five and more

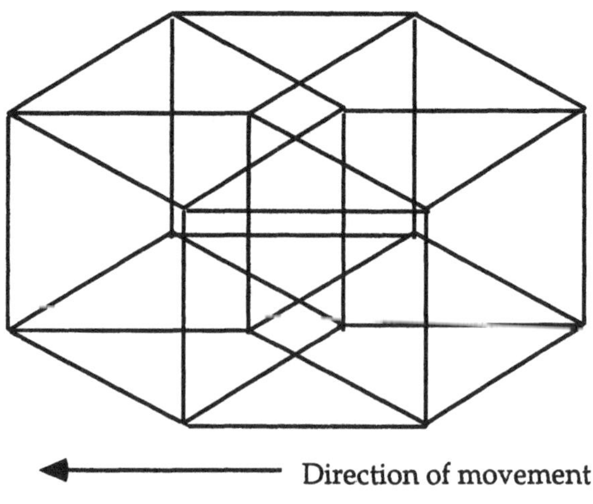

Figure 3.5

dimensions, but the complexity of graphic representation grows rapidly for every additional dimension (for examples, see Robbin, 1992). For the current purpose, four dimensions are plenty.

Just as three-dimensional space is routinely reduced to two-dimensional expression in maps and engineering drawings, four-space can be effectively reduced to three or two-dimensional representations as demonstrated above, even though these renderings can be rather difficult to read. The more dimensions there are, the more possibilities there are to deal with, and thus more gets left out in the process of reduction to a space of lesser dimensionality. Three-dimensional space, described along three orthogonal dimensions, has three possible two-dimensional normal (that is, perpendicular to each other) planes of projection and an infinity of oblique, or nonorthogonal, axes. There are then four possible planes for projecting (reducing) four dimensions of Hyperspace to the three dimensions of Spaceland. To reduce each of these further to the two dimensions of conventional graphics means that there are twelve possible orthogonally related two-dimensional permutations of a four-dimensional object. It is no surprise that observers of only one or two of these various representations would differ as much in their conclusions as do the proverbial seven blind men apprehending different parts of the elephant, or as would several Flatlanders apprehending different views of sections of a three-dimensional body passing through their plane of existence.

We are accustomed to living in a basically Euclidean, carpentered space, and to defining the world in terms of X, Y, and Z coordinates, or at least, in terms of their practical manifestation as the right-left, up-down, and in-out axes of body-space. Locally, this works, since the ground in our immediate vicinity seems basically flat, and wherever you stand on the surface of the Earth, the ground is always at your feet and the sky is always above your head. However, since we live on a curved surface—the face of the Earth—even these "absolute" reference systems will rotate as one changes location on the globe, even though each appears to be consistent in terms of human experience. An infinite number of physical locations, and therefore also reference systems, is possible. Even the two-dimensional cartographic reduction of this difficulty, made possible by postulating a uniform and global north-south and east-west reference system, doesn't quite solve the problem. The surface of the Earth cannot be evenly subdivided into squares because a double-curved surface, such as a sphere, cannot be unfolded to a flat plane without distortion. Euclidean space is essentially flat and rectilinear (square) and our world is curved (spherical)—a difference that requires fundamentally incommensurate reference systems. Translation between these reference systems is difficult, and something always seems to get left out in the process of reducing one to the other. Even though modern geometry has been able to reconcile these systems at a theoretical level, our perceptual and cognitive equipment has not yet similarly advanced.

Neither have our techniques of representation changed much in this regard. Both words and pictures are limited by their essential character as two-dimensional media (either the flat space of the page or the TV screen). Even though graphic perspective appears to provide a sense of depth, that is, a third dimension, it is only through trained observation that a reader or viewer can reconstitute this additional spatial dimension. The third dimension isn't really there. The third dimension as represented on a two-dimensional plane such as a piece of paper or a wall is only an imaginary reconstruction when it is viewed on the page, a conditioned response to a specific set of graphic or literary conventions embodied within the Flatlands of the media themselves, and which are understood only by reference to our physical experience of Spaceland and our positioning at a distance to the space of the page or picture plane, which is conditioned through our internalized knowledge of perspective. The point of view that organizes the perspectival visual field is body-centered; as such, it must be necessarily constrained by the dimensionality of the space one inhabits. The apparent distance of the observer from the observed is a convenient fiction because one can never actually step out of one's own space. There is always another point of view that includes both the observer and the observed (figure 4.14). Likewise, the convergence of lines at infinity is an illusion. No matter how long one follows railroad tracks, they remain parallel! Likewise, the diminished size of distant objects is an illusion. Perhaps the most counterproductive illusion of perspective is that the convergence of sight rays at the visual center leads one to think that one is, literally, at the center of the world; it is the illusion of omnividence, of seeing through God's eyes.

It is one thing to develop a mathematics of a one-, two-, three-, or four-dimensional space. However, the dimensions of social space may not be entirely amenable to such mathematical inscription. Along the lines of Gardner's thinking regarding the dimensions of mental space, perhaps logicomathematical expression could be taken as a single dimension of social space, and other dimensions of intelligence he identifies—linguistic, kinesthetic, spatial, interpersonal, etc.—could each constitute other dimensions. Culture, taken as whole, might then be seen as existing in, or perhaps more accurately as producing, a seven-dimensional space. It is not necessary that the dimensions of this space even be orthogonal; if they aren't, however, it complicates issues of representation in normal terms. One of the interesting facets of modern fractal geometry—the mathematical basis for many of the recent advances in computer graphics—is how it stipulates spaces of fractional dimensionality! Yet, language strains to explain three dimensions, and nearly collapses exhausted when confronted by four dimensions or by other strange notions such as that of a half-dimension. Graphics are better than language for dealing with three-dimensional spatial problems, but even graphic representations have limitations. In general we can only infer or imagine objects of greater or fractional dimensionality through great effort using existing representational methods.

There is no absolute center or origin in the Euclidean or perspectival space we are accustomed to living in. Even so, we try to center ourselves, which always leads to spatial inconsistencies. At one level, we tend to associate reference axes with local experience, and the origin with our own bodies. At another level, reference axes are based on a two-dimensional grid that subdivides the surface of the globe. North-south is aligned with the axis of rotation of the Earth, east-west is the circle of revolution, and altitude above or below sea level refers to the third axis.[8]

At an astronomical level one runs into difficulty with either of these reference systems. Astronomical distances are so large as to become meaningless with respect to ordinary experience, and, more important, the universe is not static. Everything in it is in constant motion with respect to other objects, and we have no convenient or conventional way of representing such relativity, graphically or otherwise. Not only are we not at the center of our galaxy, we are not even at the center of our solar system. Indeed, the very notion of "center" fails when confronted with the infinity of astronomical space.

In regard to human space—the sphere of everyday life, work, or politics—there is never any such ready and consistent reference as north-south and east-west. We nonetheless try to map human behavior and experience through points (words and sometimes numbers), lines (binary opposites) and arbitrary reference planes (such as right-left as a label in politics, high-low as an indicator of class, forward-backward as a level of technological development, and the opposition of center-periphery as an indicator of place in the realm of social power). But these attempts are invariably interested, relative, and partial, as any reduction of a multidimensional phenomenon through graphic, literal, or figurative projection must be.

Even though we lack formal systems for dealing with the multidimensionality of objects, processes, and events, we still manage, with varying levels of success, to cope with the unpredictability and contingency of the world through human action. However, just as our motivations can never be fully explained, all the dimensions of our actions—often intuitive, unguided by conscious intention—can never be completely recognized, any more than the results can be completely anticipated.

The challenge of thinking geometrically is formidable. Yet learning better ways of envisioning the dynamic and multidimensional character of natural and cultural space is a necessary and positive step toward better managing, or, more accurately, *designing* our own actions and how they affect the worlds in which we live. If we are to converse with the "spheres" that constitute our complex planes of existence, we must resist the limitations imposed by our reliance on the linguistic dimension as our dominant mode of communication. We need to find ways to incorporate into our everyday, conventional repertoire the multidimensional potential of other media and modes of expression such as graphics and mathematics, which are better suited to addressing spatial problems. We need, at least, to speak and write pictures and numbers as well as we have learned to do with language, and to transcend the limitations of one- and two-dimensional thinking. If we are limited only by our imaginations, by which I mean the ability to form and hold representations—image schemas—in the mind's eye, why not *cultivate* the imagination? If one cannot imagine a better world, one's actions are scarcely likely to lead toward it.

This is a challenging prospect. We all have a comfort zone, a personal world, a space of determinate dimensionality, to which we are accustomed and in which we prefer to live. Thinking geometrically—getting outside this space—may also mean living uncomfortably, and may even be intolerable for extended periods of time, since living in a space of more complex dimensionality presents certain problems and risks. Getting lost in space is always a dangerous possibility. Jason Bourne, for example, was marked as lost and targeted for execution before he was able to "come in" and resume his "real" identity as David Webb. As with Bourne, being grounded in a space of limited dimensionality may be a prerequisite for mental health, but inevitably, other problems arise that may not be resolvable within that limited frame of reference. Getting outside of one's familiar space—a form of temporary insanity, perhaps—may be the price of resolution. Getting outside of one's space also presents problems of negotiation. First, one must negotiate in the sense of navigation: One must somehow be able to get around in unfamiliar territory. But more important, one must be able to get back home. Second, one must negotiate in the sense of communication: Once home, one must be able to communicate one's findings, to persuade others who have not shared the experience to accept the vision. This is what the poor Author of *Flatland* was unable to do. As inhabitants of Spaceland accustomed to living in three dimensions, we can only sympathize and shake our heads at the rigid ignorance of his fellow Polygons and Circles. However, in dealing with challenges to our own level of dimensionality, we may be no better. To better understand the ways in which our own imaginations are similarly grounded in the dimensional limitations of conventional perception, it is time to turn to an investigation of the allegory of the Cave in Plato's *Republic*.

The Cave

The allegory of the Cave might be read most straightforwardly as a parable of the progress of the mind from a state of utter unenlightenment toward a "knowledge of the Good" (Plato, 1981/1982:227). It can also be read as an account of the limitations of perception, communication, and physical experience in respect to the development of the imagination in absolute space.

The inhabitants of the Cave have been "from childhood chained by the leg and also by the neck, so that they cannot move and can only see what is in front of them. [They] would have seen nothing of themselves or one another, except the shadows thrown by the fire-light on the wall of the Cave facing them" (227–228). Because of these constraints, they lack not only worldly knowledge but self-knowledge as well. Having acquaintance only with shadows and darkness, a prisoner of the Cave suddenly released and brought out into daylight "would be too dazzled to make out the shadows he had been used to see [sic]" (229). Our own eyes are similarly shocked by the sudden brightness of the sun when emerging from a Saturday matinee. For most of us, similarly, the mind is imprisoned in the immediate world of bodily sensation and images. In the Platonic view, we live in a state of metaphoric darkness from which it is dangerous to move too quickly. The uncultivated mind is unready for the sudden illumination of true knowledge.

The differences between inferior and superior states of knowledge parallel the metaphor of sight—especially of light and darkness—in the allegory of the Cave. The metaphor of light and darkness parallels the difference between upper and lower states of knowledge. The Cave itself "corresponds to the region revealed to us through the sense of sight" (231), which is easily misled by projected images and sensations from the "real" world. The "fire-light" corresponds to the power of the Sun, or the illumination of knowledge, which is perceived by the "soul's eye." The soul's eye alone can contemplate the "supreme splendour which we have called the Good" (232). Apart from the difficulty of defining "the Good" (or justice, the nature of which is the motivating question of the Republic according to Cornford [1981:1]), Plato illustrates the problem of getting others to agree with one's claims of true knowledge when one sees it:

> Now imagine what would happen if [the released prisoner] went down again to take his former seat in the Cave. . . . He might be required to deliver his opinion on those in the shadows, in competition with the prisoners who had never been released, while his eyesight was still dim and unsteady; and it might take him some time to become used to the darkness. They would laugh at him and say that he had gone up only to come back with his sight ruined; it was worth no

> one's while even to attempt the ascent. If they could lay hands on the man who was trying to set them free and lead them up, they would kill him. (231)

The poor Author in Flatland, yanked from the darkness of his own Space into the enlightening additional dimensions of Spaceland, was at least able to keep his wits about him, thanks to being a well-formed Square and well-versed in geometry. But when the Author was returned to Flatland, he could not convince anyone else of his sanity. Had he only known the allegory of the Cave, he might have saved himself a good deal of grief. Fortunately, the Author was spared death. In this, he fared better than his ancient counterpart, Socrates, who was put to death for bringing his views to the public. Socrates, unlike the Author, was able to recruit disciples. Thus, from the point of view of dominant interests, his actions were not only blaspheming the gods but corrupting the young minds of Athens.

Perhaps an additional reason for the magnitude of Socrates' problems is that he recognized that he had no answers, only questions. Even though "Plato believed that Socrates in his person and his life offered the real answer to the questions he propounded" he also realized that these questions were ones "which [Socrates'] teaching never solved" (Lindsay 1957:xvii). The *Republic* can be seen as an attempt to step outside the limitations of this teaching to arrive at practical answers to the questions Socrates was unable or unwilling to answer. It is relatively easy from within a frame of reference to pose questions that are unanswerable within that frame. Sometimes the only possible solution to the unanswerable questions posed from within a dimensional frame is to step outside that frame—either above, below, or sideways—and to think and act from another spatial location. This is an exercise in thinking geometrically which would doubtless be beyond the ability of any inhabitant who remained in the Cave.

The allegory of the Cave can also be read as a geometrical demonstration of the difference between levels of dimensionality. Even though the environment of the Cave is clearly three-dimensional, the prisoners in it could hardly appreciate this. Growing up in the cave not only restricts one's perception, but also one's physical experience of space. If two-dimensional projections, or shadows, constitute the whole of one's visual experience, one would lack the kinesthetic awareness gained by practice in negotiating a three-dimensional environment. However, unlike the Flatlanders, who move in a particular space where they can only see each other in edge view, the denizens of the Cave see two-dimensional images in true view, that is, as projected perpendicularly (more or less) to the actual plane of the images. The inhabitants are physically outside of the two-dimensional space of the shadows they see. Despite the distortions introduced by varying angles of incidence of lighting sources, the prisoners are always at least able to see these shadows as shapes. These prisoners live in a strange kind of in-between dimensionality, more than two but less than three. It is doubtful that any of them would be able to develop an astronomy of such a space under these limited conditions of experience.

The obvious physical problem of being dazzled by bright light is only the simplest of the problems that would be faced by a prisoner released from the Cave. Learning to see space actually takes years of practice. The experience of those who have recovered from early blindness is deeply instructive as to the ways we understand physical space and the visual character of objects (Sacks, 1995:108–152). The released prisoner, like a suddenly-seeing but previously blind person, would have no way of immediately comprehending the three-dimensionality of objects or the physical space in which they exist. He or she would lack the perceptual and verbal means to describe the sudden experience of another dimension as well as the means to physically negotiate that space without falling down or bumping into things. The inhabitants of the Cave, lacking both the physical experience of space and the logicomathematical spatial skills that derive from literacy (Ong, 1982), could not have developed, much less communicate, any concept of the three-dimensionality of their own bodies, the bodies that produce the shadows they see, or the space in which they live.

Having the conceptual tools to communicate multidimensional spatial experience may be a necessary if not sufficient condition to understand the full dimensionality of that experience, or even to have such experience in the first place. Lacking such tools may prevent any such experience or understanding. Either absolute or abstract space would likely be inconceivable to anyone who has neither writing nor mathematics with which to structure space. Even though prehistoric artifacts such as buildings, pictures (from paintings to pictograms), and carved objects demonstrate a sophisticated state of spatial awareness among our prehistoric and preliterate ancestors, we can only guess at the state of mind of people who left no other records, in particular, written records. From the point of view of an inhabitant of abstract space, the nature of absolute or mythical space is as incomprehensible as what Ong calls "primary orality" is to the literate individual.

The invention of true written script—the phonetic alphabet—in ancient Greece signaled the emergence of a distinct and powerful mode of thought: literacy, which is a prerequisite for the mode of abstract space. It can hardly be an accident that, contemporaneously with literacy, other disciplines emerged suddenly to reinforce abstract space: astronomy, mathematics, geometry, and philosophy, for example. Few of these disciplines have informed public thought as widely as literacy, however. Reading and writing are virtually essential skills for participation in everyday modern life, but no one need be a mathematician or philosopher to make a living. The functional importance of literacy in mundane affairs throughout the history of Western culture—for example in politics, business and law—bestows on literacy a value far beyond what can be assigned to other disciplines. Not even geometry, of critical importance in surveying, building, and so forth, can compete with the cultural power and flexibility of language in written or printed form, or with speech shaped through literacy, that is, with "secondary orality."

The historically subordinate status of geometry in regard to language is rooted in the mode of absolute space (as demonstrated in the *Republic*) and has persisted, though it has been transformed, in the mode of abstract space. Even though the study of several levels of geometry was a necessary part of the advanced training of philosophers in Plato's ideal city, it clearly was not as essential as literacy to everyday life even then. The fact that elementary skills such as reading and writing were not even mentioned in the description of the ideal curriculum in Book VII of the *Republic* suggests that their role and importance was simply taken for granted. The acquisition of literacy still underpins all forms of more advanced study. And the fact that its central role is still taken for granted tends to prevent a full recognition of how literacy both shapes and limits our thought and behavior.

Plato nonetheless must have recognized that the forms of thought engendered through literacy were essential to philosophy. According to Ong (1982), logic and philosophy can only be products of the literate mind. On this basis Plato's work can be read as an extended exploration of the tensions between the limitations of oral culture and the new forms of thought and social organization that were opened up by the technology of writing (Havelock, 1963). Geometry, though essential in Plato's program, was not so much an end but a means to develop these forms of thought. It was a step toward, but not quite itself true knowledge or intelligence. Plato wrote: "I think that you would call the state of mind of the students of geometry and other such arts, not intelligence, but thinking, as being something between intelligence and the mere acceptance of appearances" (226). This observation can be read in two rather different senses. First, and the sense in which Plato probably intended, is that thinking is above the stages of belief and imagining (the world of appearances), but below the stage of true intelligence or knowledge. Thinking (geometrically or otherwise)—*dianoia*—"falls short of perfect knowledge" (223). Knowledge—*noesis*—can only be attained by the higher method of dialectic. Dialectic, to Plato, is dialogue, the mutual apprehension of "an account (logos) of some Form" (223). Even though dialogue is primarily speech (as demonstrated by the fact that the entire *Republic* is written as an extended conversation), *noesis* cannot be attained solely through oral processes. The knowledge attained through dialectic (spoken or written) is transformed through literacy, which modifies oral practice. Ong (1982) calls this condition "secondary orality," that is, a state in which patterns of thought and behavior (speech, especially) have been transformed through literacy and the consequent ability to think logically. Logic is a product of the literate mind, and is as inconceivable to the preliterate mind as preliteracy is to the literate (Ong, 1967:1982). A culture grounded in orality can neither understand nor tolerate dialectic, as Socrates found out firsthand. The hierarchical difference in dimensionality of knowledge and behavior between oral and literate cultures is perhaps an insurmountable difference, as great as the hierarchical differences between Lineland, Flatland, and Spaceland.

However, Cornford observes, literate thought and dialectic alone are not sufficient to attain *noesis*. One must develop a faculty of imagination that resists description in words. Even as Plato introduces the prisoners of the Cave to represent a state of unenlightenment in contrast to the state of knowledge attained by philosophers, those who are bound by language are yet trapped by another set of chains at another level. Literacy has definite limits, which can only be perceived by getting outside of language, by seeing its limited dimensionality from a more complex dimensionality, which is an exercise in thinking geometrically. Since a more complex level of dimensionality is an infinite condition, *noesis* may be as much of an illusion as omnividence.

This condition underscores one of the internal contradictions contained in Plato's description of the several levels of cognition: Both the low world of appearances and the high world of knowledge are governed by the visual sense. For example, literacy, which is made possible by reading is impossible without vision (excepting Braille, which is probably an even less widespread practice than drawing). Without the mediating discipline of language, especially written language, the world of vision, imagination, and belief is *im*mediate and unavailable to rational scrutiny. Through literacy and literate practice, vision, imagination, and belief are mediated. The attainment of knowledge therefore involves a transformation of the act of vision—pure cognition—into *re*cognition, or conformity to existing categories and structures of knowledge. The disciplining of visual experience through literacy prevents a freer exercise of the visual sense. In a sense, we all, philosophers included, are denizens of the Cave of our own structures of knowledge. Escaping from the Cave doesn't necessarily liberate the visual sense or result in an immediate knowledge of a space of more complex dimensionality. Seeing must be learned, whether at the level of the physical negotiation of space or at the abstract level of philosophy. If one is to be a seer, one must learn to see. Plato appears to recognize this, but it is an awkward issue to deal with since the visual sense is so closely tied to the low world of sense and imagination. Confirming this paradox, Cornford observes that the highest level of knowledge, *noesis*, is "constantly compared to the immediate act of vision" (Cornford, 1981/1982:223). Knowledge and thought at all levels both shape, and are shaped by, visual understanding and intuition. The imaginative individual is always, to a certain extent, an undisciplined individual, compared to a philosopher. The training proposed by Plato is not designed to encourage imaginative thought so much as it is designed to discipline the mind. The entirely disciplined mind may no longer be free to imagine.

The second sense I read from Plato's statement about students of geometry is that thinking, being between the world of the intelligible and the world of appearances, might be understood as an *inter*mediate or transformative state of mind.[9] This is a purely modern geometrical notion. In the frame of Euclidean or perspectival reference, in absolute or abstract space, the illusory goal of either *noesis* or omnividence prevents a recognition of the validity and equivalence of alternate and partial frames of reference.

Thinking, in the geometrical sense I am after, is not only a foundational or preparatory step in the process of attaining knowledge, but a key connection, transformation, or projection between the imaginary and the apparent. This parabolic connection works in many directions. For example, thinking guided by knowledge means not only examining premises for the "ultimate principle on which they all depend" (223); it also means working out or deducing conclusions from these premises. What is required for true knowledge is perhaps not so much the formal study of geometry itself, but what it leads to: multiple ways of disciplining the visual sense, a sense that is quite resistant to description, that is, expression in words, or *ekphrasis* (Mitchell, 1994). The power and range of vision cultivated through the study of geometry (in either classical or modern form) resists being contained in the limited dimensionality of spoken or written language. Even though reading disciplines the visual sense, vision is an unmanageable slave.

Emphasizing only the formal study of geometry obscures the possibility of other, less formal, routes to geometrical thinking, which may actually be demonstrated better by the intuitive yet disciplined performance of dancers, athletes, and warriors (either heroes like Achilles, or antiheroes like the mythological Jason) than by the intellectual and theoretical work of philosophers (like Plato or Socrates). However, the antipathy of the Greek cultural elite (including philosophers) toward physical performance, especially manual work, led them to disregard the potential worth of the empirical, yet geometrical, knowledge and thought demonstrated by the performance of athletes and warriors as well as the work of less heroic (but nonetheless geometrical) thinkers, such as carpenters, masons, artists, and designers. Such knowledge has historically been disregarded as a resource in the quest for *noesis*. However, I contend that the study of how empirical practitioners develop their ability to think geometrically, albeit at a culturally lower-valued level of knowledge than that of philosophers, may tell us much more about the (social) production of (social) space than we could ever get out of the work of exceptional thinkers such as Plato or Euclid.

A deep cultural split in the Western world persists between the manual and the mental, word and image, the theoretical and the practical, not only in the scope of scholarly investigation, but in the structuring of culture and society generally. These language-based and perspectivist divisions ultimately perpetuate the privileged metaphysical status of existing social systems. These systems privilege the linguistic modes of communication and the authority of those who perpetuate the distinctions. The search for new knowledge and the solutions to practical problems will be severely limited if we remain committed to sustain the privilege(d). Geometrical thinking ought not be the exclusive province of specialists or an intellectual elite, but rather, a widely cultivated practice. As such, it is dangerous; when knowledge is no longer concentrated in the hands or minds of the few, the nature and scope of cultural power will be redefined and renegotiated at all levels.

The era Plato writes about was characterized by conflict and negotiation between traditional oral culture and the newly invented literate culture. To realize the potential of literate thought required that competitive modes of thought—including oral poetry and the visual arts—be assigned marginal status in Plato's republic. "Imagining"—both the mental activity and its physical analogue of making pictures, perhaps better termed "imaging"—is perhaps the most fundamental and necessary activity of all the (manual) arts but occupies the lowest rung of the four levels of knowledge ("cognition" in Cornford's translation). From the limited frame of reference of the Platonist, representation was imitation, a poor copy, at several removes from the essential Form or Idea. Entertaining the notion that any representation might have deeper content or meaning of its own was impossible from this point of view. The power of projection as a model for thought would not be recognized until the invention of perspective at the beginning of the Renaissance. In abstract space, the cultural role of art was drastically redefined. The experience of art became more than an aesthetic experience; it became a path to knowledge, which unfortunately still remains largely unexplored.

If the Greeks had developed true perspective, we would perhaps be further along the way toward thinking and acting geometrically. Although the Greeks possessed the geometrical knowledge to develop perspective, they never did. Ivins (1953) observes that the Greek reliance on tactile and muscular sensation as a basis for conceiving space prevented them from doing so. Still living in absolute space, though verging on abstract space, they simply had no need for perspective. The contradictions of their mode of space were, as far as they knew, resolved through language-based and logicomathematics-based means. Only when the contradictions of abstract social space emerged conspicuously in the fifteenth century would a different set of needs arise. Hence, the Greeks never gained an awareness of the possibilities for developing any discipline of the visual sense comparable in power to language-based forms of thought such as dialectic and logic.

Two millennia later, we are still under the spell these early thinkers have cast on us: the curse of logic and language. Recall Shakespeare's twist on this in Caliban's rude exclamation to his master: "You taught me language, and my profit on't is, I know how to curse!" This spell, if it persists unrecognized, may well prevent us from realizing the potential of new forms of thought that call literacy into question in much the same way that orality was challenged by script. Language is not just a tool for thought, or a medium for communication; it is an embodiment of political, cultural, and intellectual power, and a basis for socio-cultural stratification.

Thinking geometrically is both a means and an end toward remedying our cultural reliance on language-based models of thought and resolving the contradictions of abstract space. Just as literacy opened up a second dimension for the ancient Greeks, and perspective opened up a third dimension for Renaissance Europeans, internalizing additional dimensions may open up new vistas of spatial practice for us.

The development of a science of space, as called for by Lefebvre, is most likely to emerge from the application of geometrical thought to parabolic problems of transformation (especially in everyday, empirical spatial practice) that cannot be entirely reduced to words, pictures, or numbers alone. I am thinking here about representations, which are necessarily fixed and static in space. Even modern media, such as television, film, and video, that appear to show movement are as fixed in time and space as any printed document, and in some ways impose more constraints on the audience: There is no functional cinematic equivalent for speed-reading, for example. Words and numbers still enjoy considerably greater cultural status than pictures. However, as we enter the twenty-first century, the hegemony of verbal and logicomathematical intelligences has begun to wane as technologies have emerged that produce and reproduce pictures of all kinds cheaply and easily. We lack the verbal or logicomathematical means to cope with this proliferation. We also lack the means to use graphics as dynamically as we do speech in everyday life. We have no language of space, which is a prerequisite for the development of any science of space. Without such a language, we can hardly expect to apprehend the multiple dimensionality and complexity of (social) space.

A starting point for addressing this deficiency might be found in the study of drawing, and especially the little-understood genre of what I will call technical graphics, a genre which includes engineering drawing, charts, tables, and so forth. The primary language of geometry is, or at least ought to be, graphics. How we see and produce space is uniquely well demonstrated in the transformative practices and products of drawing. For several hundred years now, perspective, especially as embodied in language and mathematics, has stood as the dominant form of graphic visual representation. Even the accommodation of multiple perspectives is still a perspectival act. However, the limited dimensionality and single-centeredness of the abstract and contradictory space of perspective has now become culturally exhausted. The study and practice of nonperspectival kinds of drawing opens up new ways of seeing space. Even though there may be no way of knowing whether this is for better or worse—thinking geometrically is no guarantee of rightness—any assessment of the greater good cannot help but be better informed by having access to more complex dimensions of knowledge and experience. Life never seems to get any simpler; the development of better tools to cope with the ever-increasing complexity of the world is an ongoing necessity.

New knowledge produces new social spaces and new practices, which cannot always be reconciled with a point of view based in a space of less complex dimensionality. The contradictions of abstract space cannot be resolved from the analytic perspective of abstract space. Stepping outside abstract and contradictory space, into another more complex dimensionality, is the means for both understanding and producing what Lefebvre calls "differential" space, the synthetic space of plurality and difference. The next step toward developing a better notion of what differential space is, and how we might inhabit it, is to study the role of drawing in understanding, producing and, especially, in designing this new space.

CHAPTER 4

DRAWING SPACE

> The study of Descriptive Geometry possesses an important philosophical peculiarity, quite independent of its high industrial utility. This is the advantage which it so preeminently offers in habituating the mind to consider very complicated geometrical combinations in space, and to follow with precision their continual correspondence with the figures which are actually traced—of thus exercising to the utmost, in the most certain and precise manner, that important faculty of the human mind which is called "imagination," and which consists, in its elementary and positive acceptation, in picturing to ourselves, clearly and easily, a large and variable collection of ideal objects, as if they were really before us. . . . [W]hile it belongs to the geometry of the ancients, on the other hand it approaches the geometry of the moderns by the nature of the questions which compose it. These questions are in fact eminently remarkable for that generality which constitutes the true fundamental character of modern geometry; for the methods used are always conceived as applicable to any figures whatsoever, the peculiarity of each having only a purely secondary influence. (August Comte, *Cours de Philosophie Positive*, cited by Hall, 1903:7)

Drawing as Parable

Conceiving drawing as projection is crucial to extending its utility of the concept to other areas of knowledge, both theoretically and practically; it is particularly important in service of geometrical thinking. The historical, rhetorical, and cultural context of drawing practice and its relation to verbal and mathematical modes of communication must all be better understood in order to develop the wider potential of drawing as a formal model for thinking geometrically. One key to making these connections is, again, *parable*.

Literature, mathematics, and graphics have in common the fact that they are all graphic representations, they are typically on two-dimensional surfaces like paper, and they involve certain kinds of (parabolic) projection. For example, letters stand for sounds, numbers stand for abstract entities or concepts like "zero" or "infinity," and lines portray edges or outlines. However, as parables, literature, mathematics, and graphics each project geometric relations differently. So while language (that is, literature) has words for many of the concepts of mathematics and graphics, it is not the most effective means for investigating how projection works. In contrast with language and mathematics, drawing makes projection explicit and may be more appropriate as a basis for, and a means toward, better understanding and documenting the joint role of words, numbers, and pictures in the parabolic construction of knowledge and (social) space. This is especially true of orthographic drawing—a technical discipline based on descriptive geometry.

In this chapter my aim is to demonstrate that the joint study of descriptive geometry theory and orthographic drawing practice is invaluable in the formal—and internal—development of thinking geometrically. This is so especially at an individual level, where it offers a formal path to the development of a structured, coherent, and multidimensional imagination; but it is also so at a larger cultural level. Descriptive geometry and orthographic drawing both embody and demonstrate how multiple inscriptions can be combined to arrive at multidimensional knowledge involving individual, disciplinary, and cultural dimensions. They both provide a medium for the clear communication of the products of this imaginative knowledge, which, like all other products of human activity, are instruments and objects of negotiation, struggle, and compromise.

Beyond demonstrating its utility, it is important to understand how and why descriptive geometry developed when it did (and not earlier) and why its promise, as expressed in Comte's testimonial opening this chapter, has yet to be realized. One must thus hold the imaginative (that is, individual mental) contributions of descriptive geometry together with its socio-cultural (that is, political) dimension in what Latour (1990) calls a "binocular" view. To lose sight of either is to lose sight of the "big picture." To hold both views together is itself an exercise in geometrical thinking.

For this purpose, I begin with a consideration of early drawing practice, that is, drawing practice that precedes the development of descriptive geometry and orthographic drawing. The character of early drawing practice, like any drawing practice, must be understood in terms of multiple dimensions: its economic, technical, and material limitations and possibilities; the cultural roles it plays; the problems it sets out to solve; and the shape it gives to social space. In understanding the transformations in these dimensions, it becomes clear how and why descriptive geometry and orthographic drawing emerge amid the confluence of empirical technologies (stoneworking, timekeeping, and shipbuilding, according to Booker, 1979) and mathematical theory (in the imaginative and disciplined mind of Gaspard Monge). I then explain the basic components of descriptive geometry and orthographic drawing, highlighting their contributions as a model for thinking geometrically. Finally, I examine how descriptive geometry theory and orthographic drawing practice constitute a model for imaginative thought in other contexts, especially that of history, and speculate briefly on how the growth of computer graphics contains the dual promise to both inhibit and extend the notion of thinking geometrically.

Early Drawing Practice

The ancient Egyptians produced remarkably modern-looking line drawings on papyrus at least a thousand years before the invention of the phonetic alphabet. Surviving examples at the National Museum in Cairo show precise and consistent use of conventions such as scale indicators and reference grids, as well as top, front, and side views (or "elevations"). Despite these superficial similarities, Booker (1979) warns modern observers not to read too much into these ancient practices:

> Plans and elevations are indeed very old in principle. We must not, however, look at these in terms of our present knowledge, otherwise we get a peculiarly distorted picture. Seeing history backwards is rather different from following it forwards. Far too many people equate plans and elevations with orthographic projection, whereas it seems very unlikely that the ancients thought of their plans as projections at all. (38)

The drawings used by the ancients, despite their many similarities to modern work, were really no more than sets of instructions to be used as guides for craftsmen such as stonecutters and carpenters. Drawing was also a means of working out construction problems in the course of a project. Booker points out that much early drawing was never done on paper; rather, it was done directly on the surface of the material to be worked and would be effaced in the end. However, drawings on paper—or, in the case of Egyptians, on papyrus—might also have served as presentation drawings for a sponsor, who, much as is common today, wanted to know what a monument would "look like" before committing substantial labor and material resources to its undertaking.

The construction techniques used by the Egyptians did not require very sophisticated drawing techniques. Their buildings were based on essentially rectangular geometry. Even though the corner angles between walls sometimes varied a great deal from 90 degrees in top—"plan"—view, everything they built could be put together from blocks that sat squarely atop one another. In such construction, the sheer number of pillars required to support a stone roof prevented any development of open interior space, had they ever wanted such a space to begin with. Egyptian monumental sculpture was also essentially rectangular and looks rather square to the modern eye; hardly any of it was developed "in the round" like modern sculpture. Its rigid contours and essentially frontal presentation betray its method of construction from profiles on the sides of a block of stone.

More complex methods of projection were invented by Greek mathematicians, who apparently knew of two kinds: orthographic, which is produced by parallel rays of light, such as from the sun; and stereoscopic, which is produced by rays diverging from a central or point source of light, such as a candle. Little or no documentary graphic evidence survives from this period, but according to Booker, these techniques of projection were useful in astronomy, for example, in developing planispheres (star maps) and analemmas (figures used for correcting the seasonal variation of noon on sundials). However, the Greeks did not extend this knowledge to other potential applications of projection theory, such as the graphic solution of practical spatial problems in art and design. Why they failed to connect projection theory to graphic practice is not an easy question to answer, but Booker speculates:

> Throughout mankind's history one finds that the philosophers—today the scientists—follow up logically the work of those empiricists who have the motivation but not the rigourous kind of knowledge from appropriate fields to advance it. Undoubtedly if the ancient Greek artists had embarked on the portrayal of *things*, with geometrical shapes, rather than *people*, with complex flowing forms, a situation would have arisen in which their able geometers would have seen the problems and provided answers. (18)

No doubt the Greeks *were* somewhat concerned with the portrayal of things, because people were rarely shown without objects of some kind; I think what Booker really means here by "things" is *construction*, such as buildings and machines, examples of which are conspicuously lacking in what survives of Greek representational art. As Ivins (1946) observes, although people were a focus of their art, Greek artists did not have either the conceptual or the technical skills needed to portray the complex relations between people and things using the art and science of perspective. As a consequence, Greek art was formulaic, lacking both invention and systematic organization. Except for a few examples, Ivins argues,

> the Greek forms give every sign of having been drawn from memory of standardized objects seen from standardized positions. The Greeks struggled with the foreshortening of the individual body—sometimes successfully enough—but their foreshortening of one body had little bearing on the foreshortening of another in immediate proximity to it. The same thing is true of their notations of the shapes of objects. In general each separate body, each separate object, was seen as a thing in itself in a private space of its own. (33)

Although the Greeks understood visual angles (convergence) and foreshortening, principles that are fundamental in perspective theory, art historian and critic Panofsky (1991) asserts that ancient artists and thinkers never moved past the rudimentary and unsystematic application of these basic principles. Panofsky asks, and then answers, the obvious question: Why, given that they had the basic theoretical tools, did the Greeks never develop geometrical perspective? Indeed, perspective can be constructed using a compass and straightedge alone—the preferred tools for Euclidean geometrical proofs then as now. He answers his own question: The Greeks never developed perspective "because that feeling for space which was seeking expression in the plastic arts simply did not demand a systematic space" (43). As Ivins points out, each object, even when situated among other objects, existed in its own "private space." This quality is demonstrated in the layout of the Acropolis, where there is apparently no concern given to the relation or coorientation among the various structures; there is simply no "plan" to speak of in their arrangement, as there is in the grid of a modern city.[1]

Even though none of the antique theories of space "succeeded in defining space as a unified system of simple relationships between height, width, and depth" (43) as geometric perspective does, this lack of success ought not to be interpreted as failure. It's probably more accurate to say that the attempt was not even made to begin with. For a culture living in a different "mode" of space (Lefebvre's "absolute" space, to be precise) dominated by "antique" perspective (as Panofsky calls it, which really means nonperspectival space), the conception of any other kind of space must have been as inconceivable as the Gospel of the Third Dimension was to the inhabitants of Flatland.

Early thinkers devoted far more attention to exploring and developing the potential of literacy and literate modes of thought than they did to other technologies of visual representation. For several millennia now, the power and persistent presence of "the word" has overshadowed the potential of pictorial representation in the construction of knowledge. Plato, for example, relegated other nonliterate (and preliterate) forms of knowledge such as poetry and art to secondary and tertiary status in his Republic. The historical tension between words and pictures is based in intellectual prejudices first articulated in Book X of Plato's *Republic*, which can be read as an extended argument against oral culture in favor of a literate mode of thought (Havelock, 1963). However, according to Ivins (1946), Plato's philosophic argument against the arts may well have been based in a simple practical consideration: "[P]erhaps one of the reasons for Plato's dislike of the art of his time is its complete failure to deal with the characteristics that imply growth or 'becoming'" (24). As one of the great thinkers of all time, and perhaps the first art critic in the history of Western culture, Plato must have recognized that most Greek art, despite its substantial technical and aesthetic advances over earlier forms, simply wasn't (at least philosophically) all that good.

Greek art *was* essentially mimetic. Artists, and the consumers of art, were locked into the world of appearances, a sensual world from which Plato strove to escape. Therefore, art could serve no useful purpose in the philosopher's city-state, which was dedicated to a higher form of knowledge—*noesis*—a level of knowledge unavailable through direct sensory experience (Cornford, 1981/1982) such as vision.

Plato's criticism of the arts has little validity today despite its cultural persistence. His position must be understood to be culturally and historically specific, based in a set of cultural priorities and exigencies that no longer exist. The Greeks lived in an era of transition and conflict between traditional oral culture and the new regime of literacy, and art (like poetry) was an inevitable casualty in this struggle, due to both its technical and aesthetic shortcomings and the greater cultural priority of establishing the dominance of the forms of thought that emerged from the invention of script, such as logic and philosophy (Ong, 1967; 1982).

Greek philosophers were quick to grasp how the mind is shaped through the experience of literacy. However, at the same time that they eagerly explored the powerful intellectual and cultural possibilities opened up through literacy, they (perhaps without realizing it) closed the door on other possibilities. Unfortunately, exploiting the power and flexibility of writing (which, though fundamentally verbal, is nonetheless a specialized discipline of the visual sense) would obscure any recognition of the possibility of developing other disciplines of vision for several thousand years until the invention of geometric perspective. Had the Greeks developed knowledge of geometrical perspective, they might have gotten outside Plato's limited perception of drawing as imitative representation (*mimesis*) and thought differently about the value of art in the pursuit of *noesis*. We might then be further along the path toward gaining the "science of space" Lefebvre calls for.

"The Great Divide": The Tension Between Philosophy and the Trades

In a sense, Booker is correct in claiming that philosophers and scientists often "follow up" the groundbreaking work of empiricists, but he erroneously assumes that advances in knowledge made by theoreticians will circulate fruitfully beyond the boundaries of their own disciplines. Sometimes the obstacles are material: In the era of manuscript production, for example, the high cost of books was a severe constraint on public access to book knowledge—access that, in a print era, we now take for granted. Sometimes the obstacles are more overtly political: Disciplinary barriers to the circulation of knowledge can be traced back to ancient philosophical divisions and distinctions. Typically, scientists and philosophers throughout history have either disdained or ignored the work of empiricists such as artists, carpenters, and masons—all of whom played key roles in the invention of perspective and descriptive geometry even though they may not have known it. If ever theoreticians have attended to such work they have kept this knowledge to themselves through means such as restructuring knowledge into language that is unintelligible to anyone but another specialist.

However, the failure to share knowledge cannot be laid exclusively at the feet of theoreticians. It is due in equal part to the conservatism of trades and guilds, whose members kept their own secrets as well as anyone, and to their understandable reluctance to reject their own time-tested methods in favor of theory that might not work out in practice. If any of these "empiricists" had ever been formally trained in geometry—as philosophers were in Plato's Academy—or in other dimensions of theory, they might have been qualified to see empirical problems in a broader (theoretical) framework and provide answers on their own rather than wait for the attentions of more "rigorous" thinkers. Similarly, if philosophers had obtained experience in the empirically based trades, they might have seen both their own work, and the work of empiricists, rather differently.

To the Greeks, a work of art was only a second-hand representation, and its appreciation, being dependent on sensual experience, was deemed inferior to the knowledge attained through philosophy. Because they were tied to the world of appearances and sensual experience, manual practices were considered unworthy of philosophic attention.[2] The Platonic division between Art and Science (like the division between empirical knowledge and practice and theoretical knowledge and practice) is as much an obstacle to the construction of knowledge as is the fragmentation of drawing practices across many different disciplines within each side of the division.

Latour (1990) acknowledges the persistence of these divisions in the metaphor of "the Great Divide." Artificial divisions between fine art and illustration and between creative and technical drawing obscure and limit our recognition of how they each and jointly serve in the production of (social) space. As long as drawing and the visual arts are concerned only with the imitative or realistic representation of people and objects, their role in either the development of the visual and spatial—parabolic—imagination or the application of this imagination to other kinds of problems (in areas as varied as history and physics) cannot begin to be recognized, much less developed.

Fortunately, practical people have persisted in spite of their distance from the nuanced arguments of philosophers and other theoreticians, and technologies of drawing continued to develop along with inventions in the practical trades. The invention of the arch, for example, enabled the Romans to produce large-scale open interior spaces, which had been impossible to produce with the post-and-lintel technologies of earlier cultures. With the arch, much more complex techniques of stonecutting were required. For example, in the intersection of two barrel vaults or in the intersection of a vault and a turret, complex three-dimensional angles (known as "batters") were generated that could not be easily or completely expressed in the "standard" orthogonal views of ancient or modern drafting. These shapes were just as difficult to develop by cut-and-try methods, which, even with slave labor, would have been undesirably wasteful and time-consuming.[3] More complex technical drawing techniques were developed of necessity to deal with the spatial problems presented by the arch, although little evidence survives to tell us how early builders actually designed and built them.

According to Booker, masons, carpenters, and shipbuilders worked out their designs and building problems primarily by the means of what he calls "constructional" techniques. That is, though they worked from a design drawing, they still had to work out the actual construction details, which were shaped by constraints of the materials to be worked. To do so, they used complicated formulas and specialized tools, like adjustable squares, applied directly on the materials at hand. For example, because an architect could not anticipate the exact size of stone blocks that might be supplied, stonecutters determined the most economical use of the material themselves at the building site by working out their own angles and profiles directly on the stone. This practice was known as "stereotomy" (Booker, 1979), which involved techniques to develop precise lines on the faces of stone blocks to serve as guides for removing the excess profile—leaving a three-dimensional shape that would fit exactly with its neighboring stones. Booker provides this example (figure 4.1), redrawn from Bosse's (1663) book, which tells the reader that "there are two methods; one is to apply the bevels and a set square to the actual stone as represented in the top figure and of which I shall give no particulars" (62). Apparently, the drawing was assumed to be self-explanatory. The

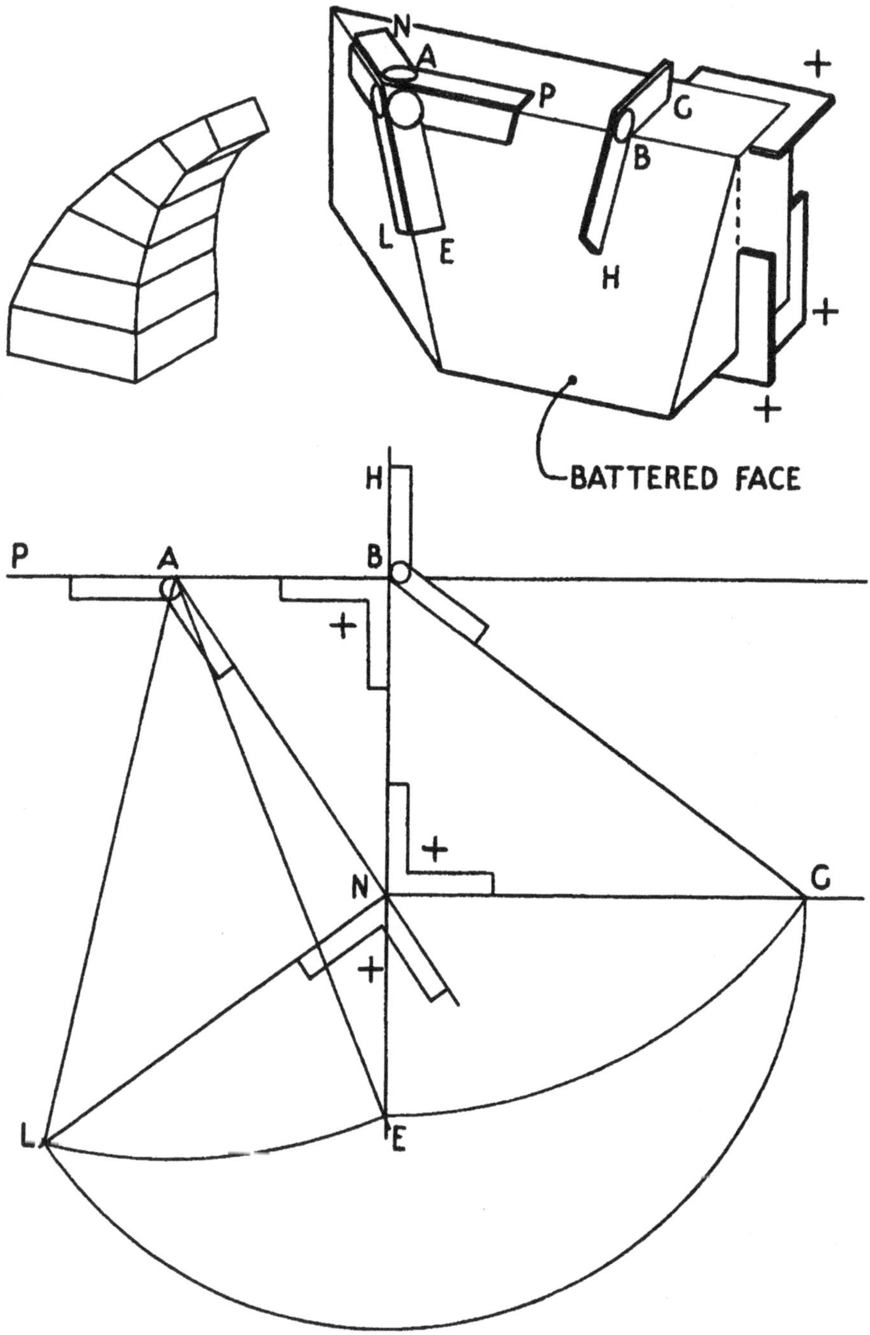

Figure 4.1

textual description of the graphic is perhaps harder to decipher than the drawing itself, but, following instructions, one ultimately finds that "the angle BAE is that of the front face panel. The angle NAL is that of the oblique side. And if you have been precise in the operation, the lines AL and AE are equal to one another" (62).

Today, sculptors pursue many of the same direct-drawing (but constructional, more so than graphic) methods for developing more complex free-form shapes like human figures. A finished carving arrived at through this process tends to betray its rectilinear origins, now as in ancient Egypt, according to the skill of the maker. Most architectural stonecutting was not, however, undertaken with the goal of arriving at a recognizable figure (as in sculpture) but with the goal of having building blocks fit together with a minimum of trial and error. The compound angles required for domes, arches, and turrets meant that many specialized tools and processes had to be developed. This knowledge was gained through practice and experiment, more often than not unguided by any mathematical theory.

Even with the publication of Masonic and other craft knowledge starting in the Renaissance, the diffusion of this empirical knowledge was slow. The conceptual difficulty and complexity of the material, its inconsistent theory, and sometimes even outright error, made stereotomy a difficult subject to master, and the essentially empirical (and formulaic) stereotomic methods were difficult to generalize to applications other than those they were specifically intended for. Developing the generalized graphical techniques of projection that would be useful for solving a wide range of spatial and mechanical problems—in the fine arts as well as in the mechanical arts—demanded a great conceptual leap. This leap was finally made by Brunelleschi when he formulated the basic rules and principles of geometric perspective in the sixteenth century (Damisch, 1994). Perspective provided a theoretical and practical basis for the joint efforts of artists and mathematicians (Dürer and Descartes, for example) who finally "changed the concept of pictures from being just representations to that of their being projections on to planes," a development that would become the "very foundation of descriptive geometry" (Booker, 1979:34).

Perspective, then, is more than a technique for graphic representation. As a model for thought it shapes the way we think and see the world. Early European art uninformed by perspective, like the art of other nonperspectival cultures, abounds in examples of the inconsistent use of foreshortening and convergence of parallels. Many Persian miniatures and Chinese ink paintings, for example, discordantly (at least to the modern perspectival eye) mix elements of both perspective and orthographic drawing in the same picture. None of the artists who produced this work could have had any sense of the geometrically unified space of perspective; they lived and worked in a different mode of space. Brunelleschi's formulation of perspective would change this situation.

The Recognition of Drawing as Projection

Despite the profound artistic and technical significance of the unified and systematic space of perspective, for the purposes of this project the most important advance enabled by perspective is that drawings came to be understood as *projections*. The primary value of understanding drawings as projections is in how they enable foresight, that is, thinking forward, especially in working out imaginary (and real) spatial and mechanical problems in things that do not yet exist, but could. This is a formal and specialized exercise of the imagination. By "imagination" here again I mean being able to hold images and relations in the mind's eye and to translate these into a systematic and precise graphic portrayal. This is something rather different from, and perhaps antithetical to, the exercise of the "artistic" imagination through perspective, as the imaginary world constructed in a piece of art need have no correspondence to relations or objects in the natural world. Until the emergence of "abstract" art, artistic drawing was for centuries (at least) dedicated to the construction of recognizable portrayals of objects (for example, portraits), in which the optically consistent application of perspective became essential. However, even in imaginary scenes, depicting what something "looks like" is essentially a projection of what has been already been seen. Despite the contributions of Renaissance artists toward publicizing the unified and systematic character of perspectival ("abstract") space, ultimately their work remains mimetic, and thus limited to the visual and sensual world of appearances.

Investigating the wider potential of drawing as a mode of (parabolically) constructing abstract knowledge would be left to the efforts of mathematicians and scientists who, through the theoretical study of perspective as projection, began to explore graphic techniques that are more abstract than representational in any strictly "realistic" sense. A Cartesian graph, for example, projects a relation between variables that often correspond to nothing that can be seen by the eye in the physical world. Geometrical concepts such as vectors, centers of gravity, or astronomical orbits are invisible; they can only be imagined, that is, visualized mentally, at which point they can be graphically projected onto paper as tables, charts, and graphs. (Maps similarly allow the inscription of mathematical information in visible form, but are, like perspective scenes, two-dimensional and reductive projections of what already exists and can be seen in the world.) Unlike the unique products of the fine arts, such as painting and sculpture, which could only be reproduced by translating them into another medium (engravings and woodcuts, most commonly) and losing much in the process, the graphic inscriptions of science and mathematics in line art were, like printed text, amenable to precise and accurate reproduction. Of course, errors can be, were, and still are just as precisely reproduced.

Once made visible, and especially when accurately reproduced, graphic inscriptions themselves are "validated" and become an important focus of research. They often become more important than their original source material. As Latour observes, working with inscriptions has several advantages over working with their source material. Inscriptions are not only mobile (portable) and immutable (they don't change over time, unlike their subject matter), they are economically reproducible, recombinable, superimposable, and scalable. More important, they can be organized into a text, and thereby integrated with established structures of knowledge: "[I]nscriptions coming from instruments merge with already published texts. . . . Through the laboratory, the text and the spectacle of the world end up having the same character" (Latour, 1990:46). Perhaps most significant of all, Latour claims, "*[t]he two-dimensional character of inscriptions allows them to merge with geometry*" (46). Even though one works with inscriptions of limited dimensionality (traditionally, on paper, but increasingly, on the computer), this information can be "made continuous with three-dimensional space" (46). In other words, inscriptions are merged not only with other inscriptions, texts, and established knowledge; they are also coupled systematically to events and objects in the real physical world through the effort of geometrical thinking.

To realize these advantages, according to Latour, inscriptions must be readable and optically consistent. Until the invention of perspective, drawings, paintings, and their reproductions were anything but optically consistent and readable. Without being based in a systematic method of projection, a drawing could be no more than a one-of-a-kind construction, devised to meet a specific situational need (whether to serve as the basis for a painting or for cutting a block of stone). This construction existed in its own "private space," unrelated to the individual spaces of other objects or drawings, except insofar as the viewer could make imaginative connections (such as those art historians and critics specialize in making). Each subsequent generation of manually reproduced copies tended to corrupt what little accuracy the original drawing might have had in the first place. Ferguson (1992) shows graphic examples of how key mechanical relations, like the direction of threads on a turnbuckle, can be misinterpreted in copying that relies on this kind of uniqueness (figure 4.2).

The graphic quality of the scan of the original drawing (B) appears to be poorer than the hand-drawn copy (A). To a casual reader, either of the drawings

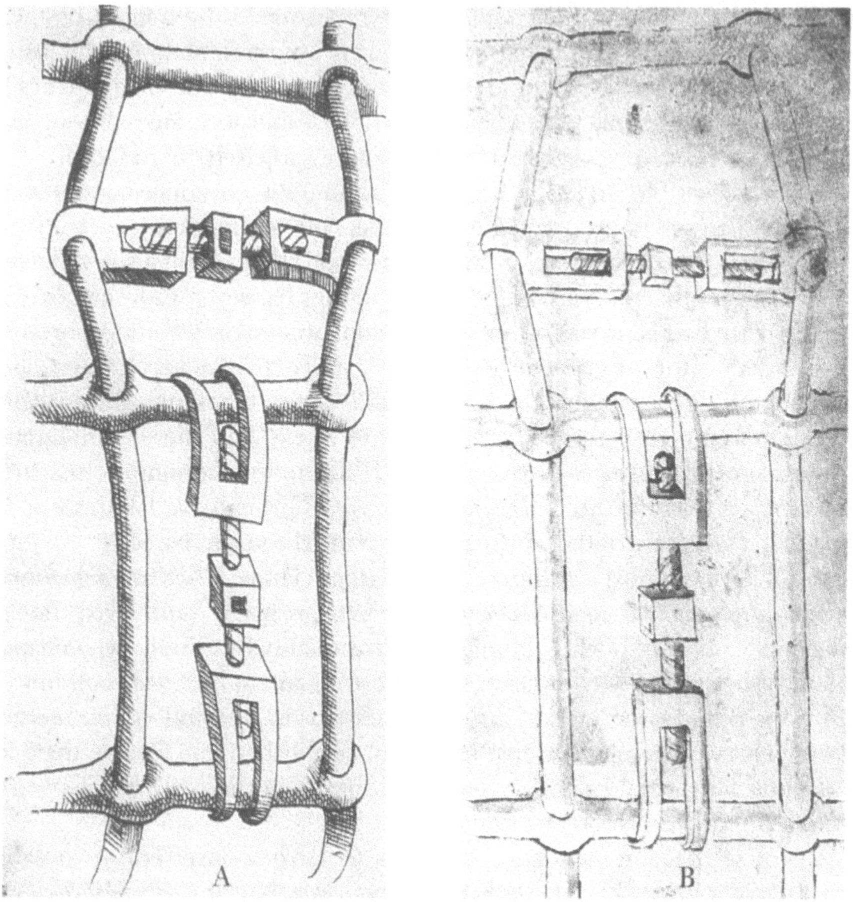

Figure 4.2

might serve equally well, but for a skilled mechanic, trying to figure out how the devices might be built, Ferguson comments that the "differences are slight but crucial. [The original] devices would work; those of the copyist would not" (113). The reason? The threads on the hand-drawn copy are reversed.

Understanding drawing as projection not only enables the production—that is, inscription—of more readable and optically consistent representations of objects and spatial relations, it also enables superior reproduction: One who can "read" a drawing copies it better than one who doesn't understand what the drawing means. Furthermore, as a model for projective thought, perspective serves as a way of extending the knowledge that it has made possible and permits clearer recognition of its own limitations. These characteristics of perspective would eventually support the invention of descriptive geometry and the formalization of orthographic drawing.

Descriptive Geometry and Orthographic Drawing

Descriptive geometry, a branch of mathematics invented in the mid-eighteenth century France by Gaspard Monge, provides a theoretical foundation for the technical discipline of orthographic drawing.[4] At the time of its introduction, it appeared to be as important as Brunelleschi's invention of geometrical perspective several hundred years earlier. However, as a subject of mathematical and philosophical fascination, descriptive geometry, unlike perspective, would prove to be short-lived. It was quickly overshadowed by the emergence of projective geometry in the early nineteenth century and the subsequent invention of non-Euclidean geometries, which, despite a lack of practical application to problems in everyday life (or perhaps even because of this lack), continue to be of greater theoretical interest to mathematicians and geometers. Today, descriptive geometry and perspective are considered mathematical cousins, branches of projective geometry that have been fully mapped out.

Descriptive geometry may be a small chapter in a closed book as far as mathematicians are concerned, but its significant role in the (social) production of contemporary (social) space has yet to be fully investigated.[5] Without the systematic and consistent means of visual imagination and communication made possible through the application of descriptive geometry theory to engineering drawing practice, modern industry as we know it could not exist, any more than mass communication as we know it could exist without the printing press.

Orthographic drawing has always been a narrow specialty, undertheorized and marginalized even in the professions, such as engineering, architecture, and others that depend on it.[6] Art historians and critics have likewise paid little attention to the mundane yet fundamental role of orthographic drawing in design. Orthographic drawing has been—and to some extent continues to be—craft knowledge that was rarely published and largely unavailable to practitioners of other crafts. Its status was comparable to the contemporary corporate practice of maintaining "trade secrets" or keeping the knowledge contained in design drawings proprietary, which restricts knowledge from both competitors and the general public. Historically, then, a variety of craft disciplines developed their own (empirical) constructional methods and applications and theoretically inconsistent systems of projection. As a consequence, different systems of orthographic projection are still in use among various disciplines and countries today.

In spite of the differences among systems of orthographic projection, the practice of descriptive geometry provides a sound theoretical basis for the coherent projection of spatial relationships in virtually any application, even if this advantage is not yet either universally understood or exploited. What these systems all have in common is that they require the translation of partial and two-dimensional inscriptions into a three-dimensional concept of the "whole." The representational value of such inscriptions is perhaps secondary to their precise and unambiguous documentation of spatial relationships, but it is a useful byproduct of their construction. Building an object from drawings does not always require a grasp of what it will "look like" when it's done, but this knowledge is often helpful. More important, though, "reading" drawings requires an accurate understanding of how the different projections of an object or concept in a design drawing are related to one another (and can be seen as transformations of each other) and "contain" information that cannot be made entirely explicit in the reduction of three dimensions into the two dimensions of a drawing. The ability to work out this translation represents an important step in being able to think in three (or more) dimensions.

The importance of graphic technologies in scientific and technical communication and innovation has been recognized for decades. For example, William M. Ivins (1953) asserts the centrality of "the exactly repeatable pictorial statement" in the emergence of Western technological culture. But Ivins, like other historians and critics of graphic communication, never really examines the crucial role of descriptive geometry or orthographic drawing in either technology or the development and exercise of the visual and spatial imagination. The much more conspicuous role of perspective in this respect has gathered far more attention (Latour, 1990; Panofsky, 1991; Mitchell, 1994; Damisch, 1994). Perspective as a formal system is not much easier to understand than is descriptive geometry or orthographic drawing, but perspective drawings are nonetheless much more intelligible than engineering drawings. This intelligibility is not necessarily inherent to perspective or human perception. Our eyes have been trained through hundreds of years of exposure to perspectival art; the apparent "optical consistency," as Latour calls it, and pictorial accuracy of perspective may be more a result of this experience than an inherent attribute of perspective itself. The ordinary citizen has little similar experience with orthographic drawings, which remain the province of specialists.

A key advantage of orthographic drawing, in comparison to perspective, is that the relations expressed in such a drawing have no necessary directionality. There is no single point of view embodied in an orthographic drawing. Any view in an orthographic drawing is easily reversible. This has certain parabolic advantages. Neither language nor mathematics[7] opens up a systematic knowledge of the multidimensionality of their subject matter in the way orthographic drawing does.

The partial and two-dimensional views of any orthographic drawing are reductions of an imaginary multidimensional object to an imaginary two-dimensional plane. These views are related to one another in such a way that they can be put back together into complete and accurate three-dimensional knowledge of their subject, even though this act requires some skill and practice on the part of a reader of such a drawing. This, above and beyond the knowledge of what things "look like" (a large part of the value of perspective), is a considerable part of the industrial and theoretical value of technical drawings in comparison to perspective renderings. As Booker (1979) observes, technical drawings are

> often used for other purposes [than representation] and, in particular, for deriving information which is inherent in the object represented but not immediately accessible. For instance, a cone may be easily represented by a circle for its plan view and a triangle for its front view; if, however, we need to make such a cone from a piece of sheet metal or paper we have to make another drawing—a development of the surface. We want an object which is implied in the representational drawing but is not given. (48)

The cone Booker describes can be "developed" or "unwrapped" only as long as the views one begins with are accurately drawn and related to one another within a consistent system of projection. Without this basis in theoretical and systematic knowledge and practice, it is nearly impossible to use the drawing for anything other than its mimetic representational content. Information "contained" yet not made explicit in the drawing can be developed accurately only by employing a systematic theory and by understanding the drawing as a projection rather than a representation. Through projection theory, one can construct a second-generation inscription that will be readable and consistent with its first-generation source.

The Need to Solve the Problem of Timekeeping

Booker claims that perhaps the earliest motivation to explore the information "contained" (yet not explicitly seen) in either the world, or second-generation graphic representations of the world, was the need for accurate timekeeping. The *clypsedra*, or water clock, perhaps the most widely used means of keeping regular time before the invention of mechanical timepieces, had to be regularly calibrated to some external reference, which in the absence of mechanical devices could only have been done by reference to regular events such as the revolution of the earth about its own axis (the day) and about the sun (the year). Even though the shadow from a vertical post appears to provide a ready means of ascertaining when the sun is at its peak (noon), a simple upright post is inadequate for consistent results all year round (because of seasonal changes in the path of the sun through the sky), and some means of compensation is necessary to determine "true" noon. Booker identifies these conditions as presenting "a very nice problem in three-dimensional geometry," which, despite the considerable astronomical knowledge of the ancients, "was not properly solved in Europe until about the sixteenth century" (50) when Desargues worked out a solution through his exploration of basic projective geometry, the fundamentals of which, like perspective and descriptive geometry,[8] can be said to be "contained" in Euclidean geometry.

However, the failure of ancient thinkers to solve the problem of "true" noon, like their failure to develop geometric perspective, is more than anything else probably a result of need and motivation. There was no more cultural need for a uniform and systematic time in Greece or Rome than there was a need for a unified and systematic space. Localized and compartmentalized measures of time and space sufficed. For example, the Romans used jugs of water that drained in three or four minutes to limit the duration of speeches in courts (Dohrn-Van Rossum, 1996); these approximate intervals had only rough correspondence to the "hours" of a sundial, which themselves varied throughout the seasons. The exact occurrence of noon, which varies throughout the seasons by several minutes either way, would have been of little concern in the practical lives of either Greeks or Romans.

Developing the methods of accurate timekeeping required for the coordination of more sophisticated and widespread business organizations in the emerging era of mass production involved not only an advanced understanding of how to solve problems of movement in space (transportation and navigation, for example, which are necessary for the distribution of goods) but also the solution of complex problems of spatial perception and representation. The issue of timekeeping itself is here perhaps secondary to the fact that, as Booker observes, advancing the "art of sun-dialling" required a new way of thinking of drawing, which, as projection, became a means of representing "three-dimensional problems rather than simple three-dimensional things" (48–49):

> [The] sun-dialling problem became of intellectual interest to mathematicians and geometers and many of these, looking at it purely as a theoretical problem, invented all sorts of geometrical techniques which may or may not have found practical application. As exercises in three-dimensional geometry they are of considerable interest, partly because they played a part in subsequent developments and partly because they demonstrate a state in which such problems were solved through a mixture of graphical and constructional techniques. (55)

The differences between "constructional" and "graphical" techniques were not significant in the technology of the ancient Greeks, Babylonians, or Egyptians. The uses for drawings were different in the mode of absolute space, as was the availability of tools and materials. For example, papyrus is a coarse and fragile medium, not suited to the repeated working—erasure and redrawing—that gets done in the course of design. The Greeks used sand trays for geometrical drawings, which are even more unstable and impermanent than papyrus. Clay, used by the Babylonians, or wax tablets, used by the Romans, have other physical limitations. Vellum or parchment is stable, but expensive. Stone is heavy, immobile, and difficult to work. Wood panels appear to have been widely used for paintings, but not much for drawings. Paper and the precision equipment used to draw on its surface only much more recently became available. Until the Chinese invention of paper reached the West a thousand or so years ago, no economical and permanent medium was available for drawing. As a result of a combination of these economic, technical, and material limitations and possibilities, a recognition of the importance of drawing as inscription and projection would not clearly emerge until the early Renaissance. Only then does drawing become something more than just a way of representing things. Only then does it become, as well, a way of identifying and solving spatial problems, and a vehicle for developing information not explicitly "contained" or represented in the drawing. Only then is there a context and need for descriptive geometry.

In summary, the slow emergence of descriptive geometry is partly due to the isolation of empirical and technical practices such as masonry, timekeeping, and shipbuilding from the theoretical pursuits of scientists, mathematicians, and geometers. Additional reasons include a lack of the necessary material such as paper and a range of precision tools. According to Pedoe (1976) the Greeks apparently didn't even have a fixed-hinge compass; they used a collapsible device instead. Such equipment makes the accurate transfer of measurements impossible. Furthermore, the fundamental problem that descriptive geometry was invented to solve—the problem of determining ballistic trajectories for optimal gun placement in fortresses—like the problem of representing a uniform and systematic space solved by the invention of perspective, simply presented itself later. The ancients had no cannons and no need to compute trajectories. Archers and spearthrowers intuitively computed and adjusted their own position and target; cannon operation was far more amenable to the calculation of trajectories. The difference between modes of space—absolute space and abstract space—in each era was ultimately connected to the practical, empirical conditions and problems faced by soldiers and their leaders. Descriptive geometry could only be a product of someone thinking and living in abstract space; the uniform and systematic graphic representation of objects and spatial relations enabled through descriptive geometry would have been inconceivable to anyone living in absolute space.

The Emergence of Descriptive Geometry

The story of how descriptive geometry came about, how it was received, and what happened to its inventor echoes both the story of Flatland and the allegory of the Cave. Gaspard Monge, the inventor of descriptive geometry, plays a central role in the story. Monge was probably unique among his contemporaries because of having received formal training in mathematics and geometry as well as having firsthand experience in the three craft disciplines Booker identifies as key ingredients for the invention of descriptive geometry: stonecutting, timekeeping, and shipbuilding. Without this combined experience—a mix generally unavailable to scholars of his time—it is unlikely that he could have arrived at his invention.

Monge entered the school of military engineering at Mézières in the early 1760s. As the son of an itinerant Savoyard merchant, he was admitted as a draftsman rather than as a regular student. Confronted with the tedious task of calculating optimal gun placements in a proposed fortress, the 16-year-old Monge developed a graphical method that enabled him to solve such problems in a fraction of the time required by the conventional arithmetical procedures he was expected to use. At first his superiors refused to believe what he had done, but he was able to prove both his method and its results. He was subsequently allowed to teach his method, which he named descriptive geometry, but he was denied the opportunity to publish his work. According to his biographer Aubry (1954), a combination of the conservative resistance of trade guilds (stonecutters and carpenters in particular) to his method and the government's recognition of its military advantages resulted in its suppression as a state secret for over 30 years.

Monge was a man of formidable intellect who embraced a wide range of humanitarian and scientific interests above and beyond the invention of descriptive geometry. He was active in revolutionary causes and was much concerned with the inferior political, economic, and technological status of France in the early days of the Industrial Revolution. Monge felt that an improved system of education was necessary to close this gap. He was elected to the Académie des Sciences at the age of 34 and was instrumental in the founding of L'Ecole Polytechnique in the 1790s, where he was finally able to lecture and publish his memoirs on descriptive geometry.[9]

During this time, he also became a close friend and confidant of Napoleon. After the restoration of the Bourbons, however, he was discredited as a Bonapartist and stripped of his membership in the reorganized Académie in 1816. At about the same time, the curriculum at L'Ecole Polytechnique was restructured and his educational ideas lost out to those of Lagrange, who favored "a more traditional and theoretical diet of advanced calculus and rational mechanics" (*Encyclopedia Britannica*, 7:582), a philosophy which continues to dominate most engineering programs to this day. Monge died in 1818. Today, the life and work of Monge is little known except through a brief, but nearly universal, citation in introductory texts on descriptive geometry (traditionally a mandatory basic course for engineering students), which describes him as "the father of descriptive geometry."

Monge must have been deeply frustrated by the opposition to, and limitations placed on, his work, both at the start, when he was opposed by the entrenched conservatism of guilds and traditional apprentice programs and restricted by the government, and later, when he was finally undone by the established joint interests of military, political, and academic elites—the very interests that had the most at stake in maintaining the production of abstract (perspectival) space. These interests, which ironically both motivated the invention and the initial suppression of descriptive geometry, along with the widely acknowledged difficulty of the subject, limited its application to other areas of knowledge. Beginning in the social space of the early Industrial Revolution and continuing today, the theoretical potential of descriptive geometry as a tool for developing, disciplining, and exercising the imagination—as a means of learning to think geometrically—would be overshadowed ultimately by its own practical utility, as well as limited by the priority and dominance of perspective as a model for thought.

Even after being published and taught, knowledge of descriptive geometry was slow to spread. The conceptual difficulty of its subject matter was undoubtedly a substantial obstacle, but, perhaps even more important, the ancient and persistent divisions of manual and intellectual labor and art and science played a crucial role in limiting the development of the imagination through descriptive geometry. (Scientists and mathematicians deal with abstract problems, and builders deal with the concrete!) Were theorists and empiricists to find a common language, the knowledge of each, and culture generally, might be considerably advanced through the experience.

It may be too much to expect that established disciplines and interests would at any time willingly dismantle the boundaries that separate their partial and compartmentalized knowledges. At pivotal moments in history, however, certain individuals—geometrical thinkers, all—have performed this service. Although the contributions of Monge to the production of the space of modern technological culture have never been as widely acknowledged as the momentous discoveries of these other more conspicuous figures (Brunelleschi, Galileo, Kepler, and Newton come to mind), his work deserves reexamination as a model for thought; descriptive geometry is more appropriate than perspective as a means toward the production of differential space.

The Contribution of Ship Building in the Development of Descriptive Geometry

Monge was undoubtedly familiar with stonecutting and building practices, but it is hard to guess whether he, as much as Desargues, ever became aware of, or involved in, problems of timekeeping. At any rate, these problems were already solved, and the invention of accurate clocks had made sun-dialling a largely obsolete pursuit. And Monge, like any ambitious youth, was looking for new problems to solve. Interestingly enough, he was involved in the Napoleonic project to develop a decimal calendar in the late eighteenth century, a social experiment which was actually adopted, but shortly thereafter abandoned. More important, however, early in his professional life as a naval examiner he came into contact with the practices of ship design and building. His exposure to these disciplines had to have had an important effect on both his generalized method of descriptive geometry and his later contributions to differential geometry, which deals with the mathematics of generated surfaces, such as the hulls of ships.

Buildings and sundials are fixed in space and time, and despite the difficulties presented by some of the problems involved in their design and construction, both are relatively simple compared with the design and construction of ships, which move freely in three dimensions and must remain stable under all conditions: from empty to fully laden, from calm waters to heavy seas. Ships, therefore, present both designer and builder with a considerably more difficult set of spatial and constructional problems than almost any kind of architectural or timekeeping problem.[10] Developing the intersection of two single-curved surfaces like half-cylinders, as in the case of two-barrel vaults, is simple in comparison to the problems involved in dealing with double-curved surfaces like the hull of a ship. For optimal performance, ship hulls are made of freeform (spline) curves that cannot be completely expressed as conic curves, like circles, parabolas, hyperbolas, and ellipses, which the Greeks investigated early on but which were not fully understood until projective geometry was fully developed in the nineteenth century by Poncelet.[11] In practice, however, spline curves can be closely approximated by a series of radii, which are relatively easier to draw than conics with simple tools like pivoting dividers and compasses and flexible curves (tools not possessed by the Greeks; one can only wonder how they went about designing their ships). Renaissance shipbuilders developed accurate, empirical techniques for locating and drawing the shape of a hull. This was usually done at full scale on the floor of a shop, a practice known as "lofting." The curves developed were then used as templates for constructing the individual pieces (ribs and planks), which, if accurately shaped, would fit perfectly together in the finished hull.

Typical wooden ship construction starts with the keel, which is at the lengthwise center of the hull. Symmetrical ribs of gradually varying size are placed crosswise to the keel at regular intervals. Both keel and ribs are fairly simple to develop because they curve in only a single plane. Planks were then affixed to the ribs, and the natural springiness of wood ensured the smooth development of curves along the length of the hull. However, the lengthwise shape of these boards, before installation, is never straight. Because of the differences in height and width between bow and stern, these boards not only bend; they twist, forming a compound curve. Their proportions and edges must be accurately determined in advance if they are to fit tightly together to prevent leaks in the finished ship. Without an accurate means of predicting the end result, considerable time, material and, therefore, money would have been wasted. Complex drawing methods were developed to ensure the correct proportioning of these planks to the change in curvature of the ribs, so that no bulges, flat spots, or gaps would develop.

Fit, finish, and appearance, however, are not the final determinants of success in shipbuilding. A ship must float and move efficiently through the water under many different conditions. A fully laden ship will sit lower in the water than a partly loaded or empty ship. The curve of the waterline must be accurately anticipated to achieve stability in all states of loading. According to Booker, making the imaginative connection between this real condition—the variable location of the waterline—and the notion of that line being a theoretical "section" through the ship would serve as the basis for more sophisticated methods of design drawing.

To accomplish this connection, generated surfaces are expressed graphically as sections. A section is the result of a surface intersecting a flat plane. After realizing that a waterline is a section, any number of such horizontal sections can be developed, and these are extremely useful in describing the complete shape of the hull. Extending the concept, the keel and the ribs, similarly, can be seen to represent lengthwise and crosswise sections of the hull, as Booker (1979:71) shows in his example of a pilot boat design drawing from about 1900, reproduced in figure 4.3.

Although the actual shape of any hull plank usually does not conform to any particular X, Y, or Z section, these actual curves, called "flowlines," can be accurately extracted, or developed, from a set of X, Y, and Z sections, which "contain" the information, though not explicitly. Virtually any freeform (curved) body can be similarly expressed as a series of sections, even though

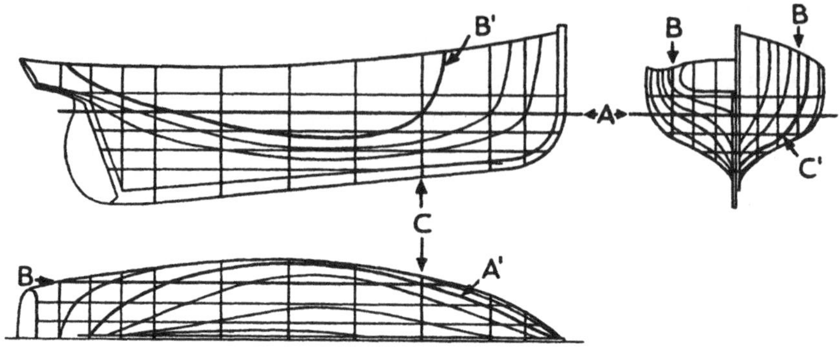

Figure 4.3

this is time-consuming and tedious for complex shapes such as the human body. Today, various scanning devices perform this work: Modern technologies project three-dimensional shapes as a series of sections, and this information can, in turn, be used to reproduce the shape accurately through stereolithography.

Monge realized that a series of sections is also a way of expressing movement through space, as was illustrated in chapter 3 in the example of the Sphere passing through the plane of Flatland (figure 3.1). Through the imaginative exercise of descriptive geometry theory, the partial and two-dimensional information represented in both sections and projections can be combined into an intuitive but accurate and systematic three-dimensional knowledge of complex objects. However, some complex details of curved surfaces—such as are found in carriage, automobile, and human bodies—are not always as easily reducible to expression in X, Y, or Z sections, as are the hulls of ships, and must be carefully developed in the design process through a series of what are called "auxiliary" views or sections.

In manufacturing, for example, the curve of a molding is generated by the shape of a cutter (such as an S-curve, or ogee) as it is moved through space. The "true" shape of the molding is defined by the cutter as well as by the direction of movement. This three-dimensional curve cannot always be easily generated on paper. It is often time-consuming and tedious to project such a shape accurately, but it must be done to make sure that parts, such as a door and its frame, line up properly when assembled together.[12] More complex surfaces, such as those generated by a section that changes shape as it follows a complex path in space, are still more difficult to develop graphically, but this too can be done. Contemporary methods of solving such problems are based almost completely in descriptive geometry theory, which provides a consistent theoretical basis for not only developing, but integrating, the partial representations of sections and projections into a complete and coherent document.

The Legacy of Gaspard Monge

Monge was a geometrical thinker of the first rank. If for no other reason than his invention of descriptive geometry, he ought to be considered one of the pivotal figures in the emergence of modern Western technological culture. Without his contributions, modern industry and technology might not exist in the form they do today. However, the identification of his work with the obscure industrial practice of "drafting" has diminished the recognition of the fundamental importance of descriptive geometry and orthographic drawing in all dimensions of contemporary culture. Often credit that should be extended to Monge is displaced. For example, Ivins (1953) identifies the print—the "exactly repeatable pictorial statement"—as a crucial factor in both the development of modern science and the Industrial Revolution. But the accuracy of prints, and the design and manufacture of the precise machinery that enabled their efficient reproduction, would not have been attainable without the widespread industrial application of the graphic "language" Monge invented, in use now for over two hundred years.

Examining all the reasons why Monge and his work are not more widely recognized is a project beyond the scope of this book.[13] One additional reason, beyond the largely political and disciplinary reasons discussed above, is, however, particularly relevant here. As Booker (1979) puts it: "It seems that the then academics misinterpreted the nature of descriptive geometry [and] imagined that they had a *really new* geometry in their hands which could be used to extend our knowledge" (132; italics added). Perhaps these academics, like the allegorical denizens of Plato's Cave, were temporarily blinded by the sudden light and novelty of descriptive geometry and misunderstood its significance. As it turned out, their senses would soon recover, and they would regain their "perspective." The theoretical promise, and novelty, of Monge's invention was exhausted rather quickly, and the academic fascination with it was soon eclipsed by a greater fascination with the invention of projective geometry by Monge's student Poncelet, which not only subsumed both perspective and descriptive geometry, but pointed the way toward the subsequent invention of non-Euclidean geometries in the middle and late nineteenth century. These geometries, however, although more theoretically fruitful than descriptive geometry, are both less intelligible and practical.

Among mathematicians today, Monge is recognized more for his pioneering work in the field of differential geometry (the application of calculus to the development of curved surfaces) and analytical geometry (the algebraic solution of three-dimensional problems) than he is for inventing descriptive geometry, which is largely seen as an empirically rooted, manual, and graphic approach to problem-solving that bears little connection to the work of theoreticians in mathematics and science. For example, Booker observes that even though "Monge's descriptive geometry appeared to be very important at the time it was introduced . . . over the intervening years its real worth has become more and more difficult to assess. This assessment is, today, probably very much a matter of opinion and depends largely upon the kind of drawing education one has had" (86). Indeed, contemporary drawing education and practice still tend to be empirically based and emphasize what Booker calls "secondary" geometries (that is, practical formulas or rules for solving spatial problems)[14] rather than being conceived of and taught as a means for developing an understanding of the "primary" geometry of projection theory itself. In this setting, the primary knowledge of descriptive geometry is considered an unnecessary level of "overlearning." In most engineering programs, the complex theory of descriptive geometry—the mastery of which was once practically a rite of passage for engineering students—is now represented only vestigially.

In some ways this is understandable. Descriptive geometry might not have been "really new" after all, but was an important—of perhaps greater significance than its evanescent theoretical fecundity—systematization and unification of the previously inconsistent, compartmentalized, empirically developed knowledges and practices of earlier forms of orthographic drawing. Its intimate and fruitful association with empirical and manual practice and knowledge—industrial applications—has had the unfortunate consequence of making descriptive geometry rather unappealing to "pure" mathematicians and geometers, who feel no need to relate their research to problems in the real world. The practitioners of orthographic drawing, on the other hand (as Booker observed), even today generally lack the "rigourous kind of knowledge" or theoretical background to extend their knowledge beyond the immediate problems of their profession. In this, we really haven't come very far since the era of the Greeks.

The accurate representation and documentation of three-dimensional things, such as machines or buildings, is a largely practical pursuit and of much less interest to theoreticians than it is to practical men and women. However, the value of descriptive geometry is that it enables the coherent representation and solution not only of objects and relations in space, but of multidimensional problems as well. This is an important theoretical pursuit that transcends other more mundane applications of descriptive geometry. For example, determining the center and radius of a sphere that passes through four noncoplanar points in space, an unwieldy problem to solve by any other method, becomes relatively easy through the application of descriptive geometry. Though stated here as a theoretical problem, it is a problem that might be encountered by a designer trying to determine how much space is available inside a complicated set of mechanical components, such as in the engine compartment of an automobile. By a shift in definition of the subject matter of such a problem, it is possible to generalize the utility of descriptive geometry to other kinds of spatial problems that are not of a mathematical nature. For example, if one were to (parabolically) conceive of points or lines as representing abstract entities or concepts, rather than things, and their frame of reference as social rather than physical or mathematical space, it might be possible to graphically represent—to make visible as inscriptions—such invisible phenomena as changes in social structures, relations, and forces, much as Cartesian graphs make visible the abstract operation of invisible physical relations and forces such as gravity and acceleration.

Understanding drawings as projections that are not necessarily perspectival—or even representational—in character is a key to opening up their potential as a (parabolic) model for imaginative thought and developing their utility as a tool for investigating spatial problems in disciplines that are not as specifically visual as design and drawing. In language, for example, metaphor is a way of making an imaginative (and parabolic) connection between a concept and a verbal representation (usually, a word). Both the act of projection and the metaphor of projection play a substantial role in the exercise of the visual and spatial imagination through language.

Culture—including history, politics, economics, and so on—generally constitutes a far more complex and multidimensional space than we have the means to understand or to represent with any real accuracy at present. Perhaps, through developing a science of space and a suitable "language" for projecting space that would serve as powerfully as perspective once did, such multidimensional problems might be amenable to solution. Despite the advances made by artists and scientists, we are now about as far along in developing the conceptual and technical skills for such projects as were the Greeks in representing the equally problematic relations (for them) between people and/or things in space.

The development of systematic methods of both perspective and orthographic drawing as consistent and reliable means of documenting and solving problems of space was a project stretching over multiple disciplines and several hundred years; building the geometrical knowledge to represent and solve even more complex spatial problems (such as economics and history present to us) may be a project of many centuries to come. Descriptive geometry, as the theoretical basis for most contemporary technical drawing, can provide the basis for extending the geometrical imagination in the solution of all kinds of problems of space, including the production and inscription of "differential" space, and perhaps even the construction of the "quasi mechanics of history" sought by Ortega y Gasset, which may be the same as the "science of space" that Lefebvre calls for. The scope of this chapter does not permit giving a full account of descriptive geometry, which, at any rate, like complex theory in other disciplines, cannot be fully grasped through reading alone, but must be worked out through exercise, experience, and practice. A basic knowledge of how the subject is embodied in the basic theory and practice of orthographic drawing will nonetheless suffice.

Orthographic Projection

Any three-dimensional shape in space can be depicted by a series of lines, which typically indicate outlines and edges. Perhaps the simplest way to produce an accurate drawing such as a profile is by tracing a shadow (a projection). However, the various possible projections that can be made from different viewing angles all result in different outlines. Without some underlying system of recreating projections there is no obvious means of consistently relating these different outlines to either the object from which they are generated or to other views of the same object.

Monge's basic insight was to realize that the exact location or shape of any line, or any point on a line, projected onto a plane in three-dimensional space can be exactly determined by reference to a second plane of projection if this plane is orthogonal (perpendicular) to the first. In the simple case of a straight line in space (AB), this is graphically demonstrated by the figure on the left in figure 4.4.

Through the interrelation of parallel rays of projection (AA_1, BB_1 gives projection A_1 B_1; and AA_2, BB_2 gives projection A_2 B_2), and the fact that the two

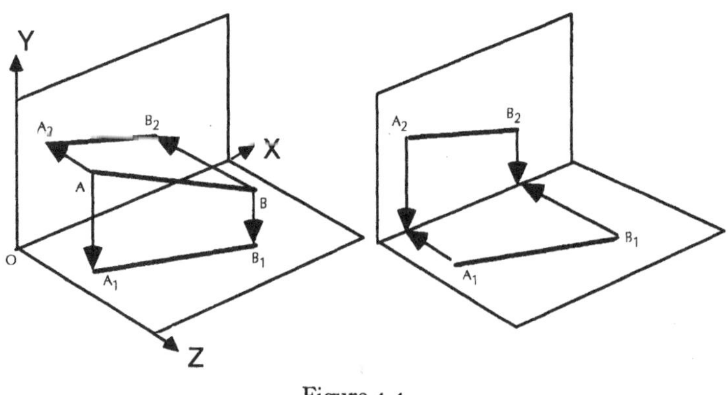

Figure 4.4

planes (XOZ and XOY) are orthogonal, the two views can be accurately related to one another. Note how the two projections meet exactly when they are projected orthogonally onto the axis of the intersection of the two planes, as shown by the figure on the right in figure 4.4. By mental effort or graphic construction, following the rule that views are orthogonally related, the three-dimensional entity shown partially in each view can be reconstituted.

Most, but not all, basic principles of Euclidean geometry are preserved in this method. For example, straight lines that are parallel in three dimensions remain parallel in their projections. However, curves, such as concentric circles, do not necessarily remain parallel, though they remain concentric. Angles are foreshortened. In any case, for any object or set of relations, it is possible through this method to construct a view, or series of orthogonally related views, that shows the "true" relation of any set of lines in space. It makes no difference whether these lines represent a real object, such as a carving to be made in stone, or an abstraction, such as a ballistic trajectory, or a theoretical problem, such as a curve that graphically depicts acceleration.

It is this generality and flexibility in translating objects and relations—whether real or imaginary—into visible inscriptions that make descriptive geometry a powerful tool for the imagination. The application of descriptive geometry to design practice brought a new theoretical rigor to the problem of seeing, imagining, and representing objects and spatial relations in a coherent and nonperspectival framework.

Since no one orthographic view can show all the details and features of even the most simple objects, multiple views are usually required. These views are taken at 90-degree—"orthogonal"—angles to one another. There are six theoretical possibilities in the 360 degrees of three-dimensional space: top (plan), bottom, right, left, front, and back views. Usually, only two or three of these standard views are sufficient. Figure 4.5 shows the example of a house in plan, front, and side views.

Through this technique, a single object, no matter how complex, can be accurately depicted, though not always in detail sufficient for someone to actually build the object from these few views. However, such an object always stands alone, in its own internally constructed reference space, and cannot necessarily be related accurately to other objects, for example, to other nearby houses or landscape features. Descriptive geometry theory provides a unified frame of reference for multiple objects by relating views to the X, Y, and Z axes of Cartesian or abstract space. In the case of a single object, for convenience these axes are usually drawn along major features of the object, as illustrated in figure 4.6.

Technically, these axes are not just lines. They represent the edge views of three theoretical planes of projection. Mathematically, each plane is usually

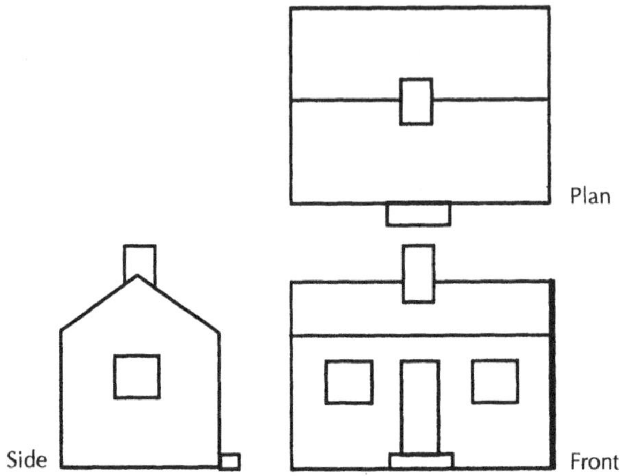

Figure 4.5

defined by the two axes that are seen in a particular view. Thus, the plan view is located in a X-Z plane, the front view is located in an X-Y plane, and the side view is located in an Y-Z plane.[15] In each two-dimensional view, the third dimension cannot be shown. The "missing axis" defines the view direction: The plan view can be conceived as a view along rays parallel to the Y axis, the front view as a view along rays parallel to the Z axis, and the side view as a view along rays parallel to the X axis.

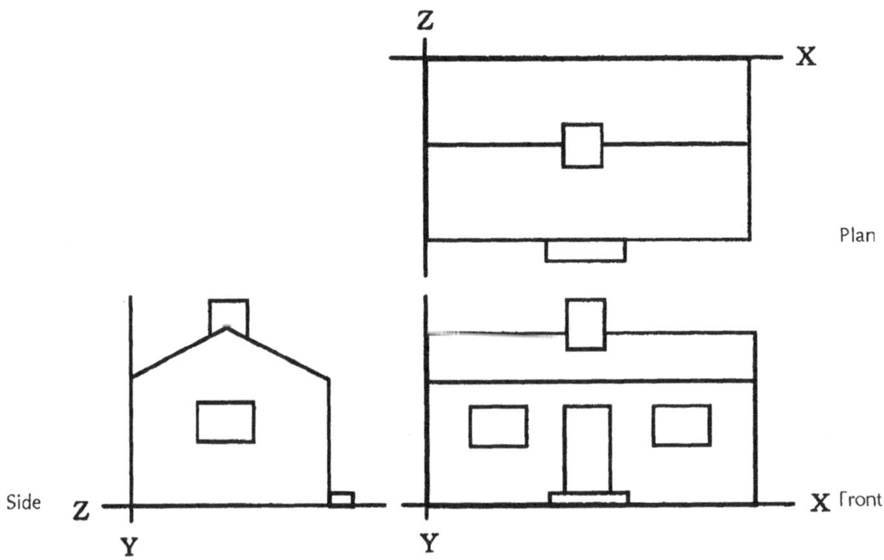

Figure 4.6

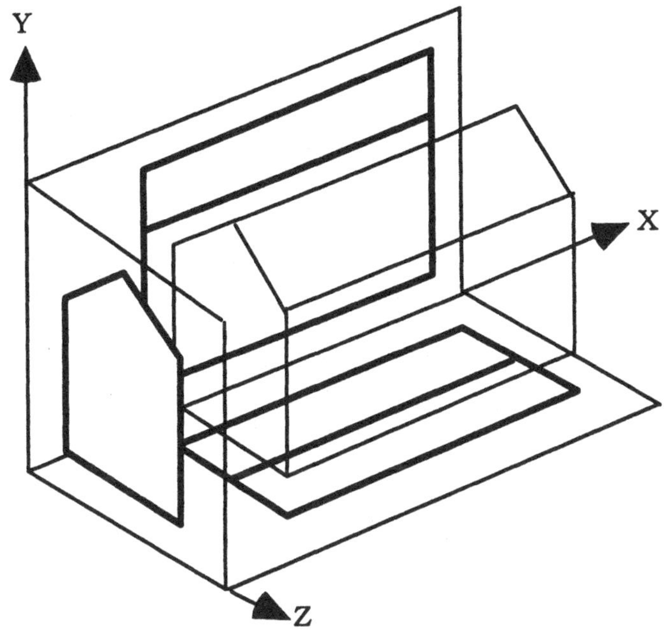

Figure 4.7

The actual arrangement of these views, their relative priority, and the labels associated with each differ considerably according to disciplinary and geographical boundaries even though most of the basic theory behind different systems of drawing is much the same.[16] In the United States, "third angle" projection is standard (figures 4.5 and 4.6 follow this convention); in continental Europe, "first angle" is dominant. Belofsky (1991) refers to these as differences in "dialect" and asserts that they are rooted in political, economic, and cultural interests more than in technical needs. However, these differences have little significance in regard to the utility of technical drawing as a tool for the imagination. Both systems are theoretically equivalent. Although third angle projection offers certain practical advantages, first angle projection offers certain theoretical advantages; these will be discussed briefly later.

Because there is no absolute reference or origin point (zero X, Y, and Z) in physical space, the "origin" of the axes is located through convenience and convention. In the case of multiple objects, in a group of houses, for example, the reference coordinates might be located with respect to landscape features, such as an imaginary ground plane (elevation above sea level) and north-south, east-west axes.

Although all these axes appear graphically as lines, it is important to realize that because an edge view of a flat plane is always a straight line, these lines are actually representations of theoretical planes. In descriptive geometry, each "view" of an object is conceived as a projection of an object in space onto the imaginary plane constituted by any two axes. This can be shown clearly only in a "rotated" view,[17] which shows all three planes of projection at once. This is demonstrated in figure 4.7, where for maximum clarity details such as windows and doors have been eliminated and the object has been moved away from the reference planes.

In this arrangement, the projections might best be conceived as shadows of the actual object, as though the object were transparent or made of wire and light rays for each view were cast perpendicular to the particular plane of each view.

If these planes were conceived as three sides of an opaque box, hinged along each axis, the box could be unfolded so that all views lie in a single plane, typically, with the plane of the major view, the designation of which varies among disciplines, at the center. In figure 4.8 the major view is the front view.

Notice that the "top" view here appears *below* the front view, and the right "side" view appears on the *left*. To the empirically based practitioner, this is counterintuitive: Shouldn't a view from above be located *above* the plan view, and shouldn't a right side view be located on the *right*?

This brings us to the fundamental difference between first angle projection and third angle projection. Figure 4.8 is actually an instance of first angle projection, in which the object is located in what is called the first quadrant, or angle, of Cartesian coordinate space, as illustrated in figure 4.9.

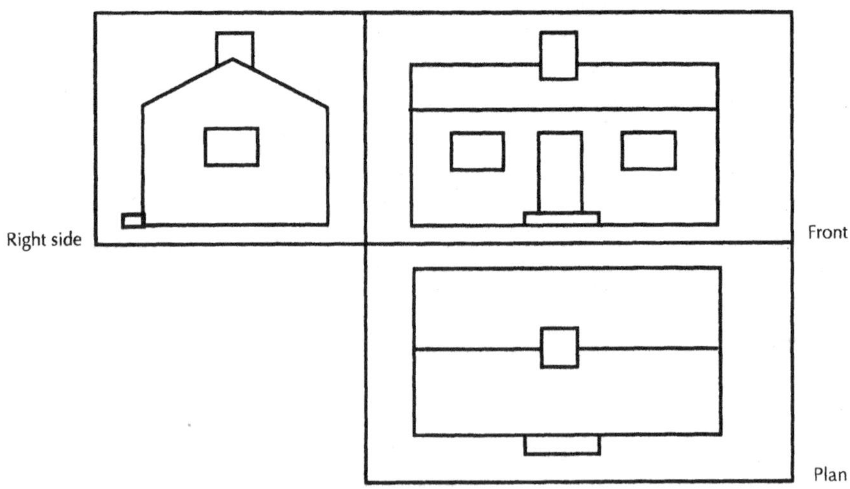

Figure 4.8

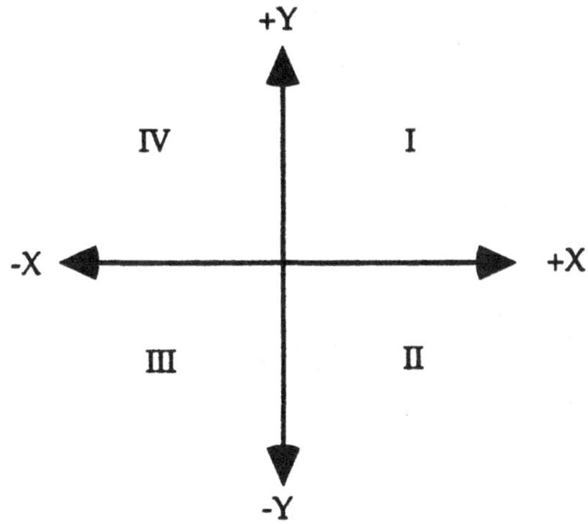

Figure 4.9

The first quadrant, because it shows positive values for both X and Y, is conventionally preferred for graphs. The Z axis is here the "missing" axis, and passes perpendicularly out from the page, through the intersection of X and Y. For various technical reasons that are not important here, the second and fourth quadrants are not useful for projection.[18]

Third angle projection, as illustrated by figure 4.10, is quite useful. In this system, one must envision a transparent (glass) box on which the traces of the object are projected by imaginary rays perpendicular to each plane.

Unfolding this box is rendered in figure 4.11, which matches and now makes sense of the example shown in figure 4.6.

Monge was familiar with and must have been inspired in part by the practices of stonecutters in Mézières whose drawings on the faces of a stone block as preparation for carving are actually instances of third angle projection. But he must have realized that basing his work on empirical (craft) practice, rather than the established theory of Cartesian coordinate space, would be a questionable move from a theoretical or academic point of view, which would favor first angle projection. Unfortunately, a large part of the resistance of the guilds to his invention may have rested on an objection to this choice, as much as it rested on objections to its counterintuitive form, its practical complexity, and its potential for error in translating first angle into third angle projection.

This discussion is not the place to examine the full range of these issues, which have been investigated at length by Belofsky and Booker. Suffice it to say that, even though Monge developed a systematic theory to unite the disparate and often inconsistent empirical practices of technical drawing, craft

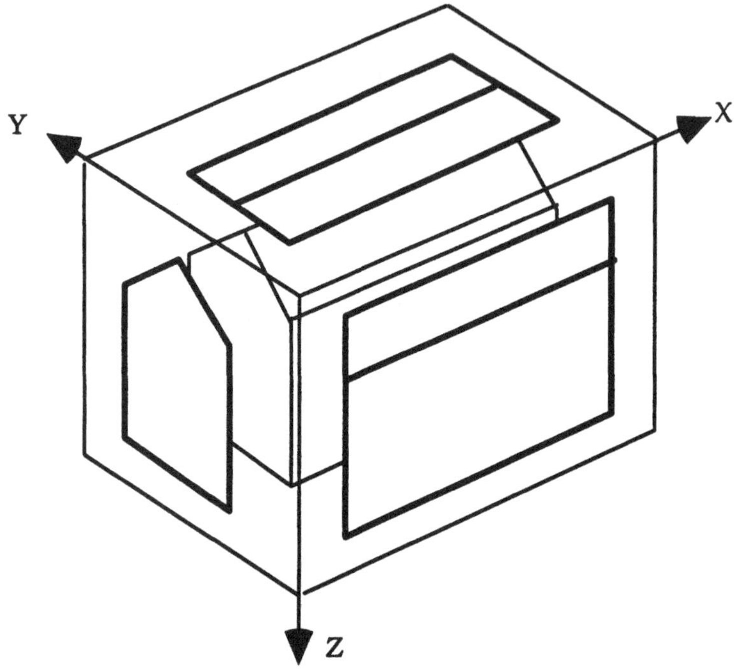

Figure 4.10

practitioners for many years largely ignored the potential of this theory in favor of their own traditional methods. Likewise, theoreticians, despite a brief flurry of fascination with the elegance of his invention, quickly moved on to other more novel and promising areas.

The gap between theoreticians (such as philosophers and scientists) and empirical practitioners (such as craftspeople and artists) has been characterized for millennia by prejudices on both sides. At the same time that theorists have disregarded the importance of the achievements of empiricists throughout history, the empiricists have equally disdained the importance of theoretical work that might enhance their own practical ability. Although Monge's invention of descriptive geometry must have been partly inspired by his early exposure to, and experience with, the craft knowledge of stonecutters, carpenters, and shipbuilders, his professional and cultural allegiances and career interests sided with the cultural elites. In the end, he was disenfranchised from all these connections.

A large part of the value of descriptive geometry rests in how the fundamental principle of orthographic projection onto perpendicularly related planes that are "square to the world" can be extended into oblique and multiple frames of reference. This is done through the development of what are

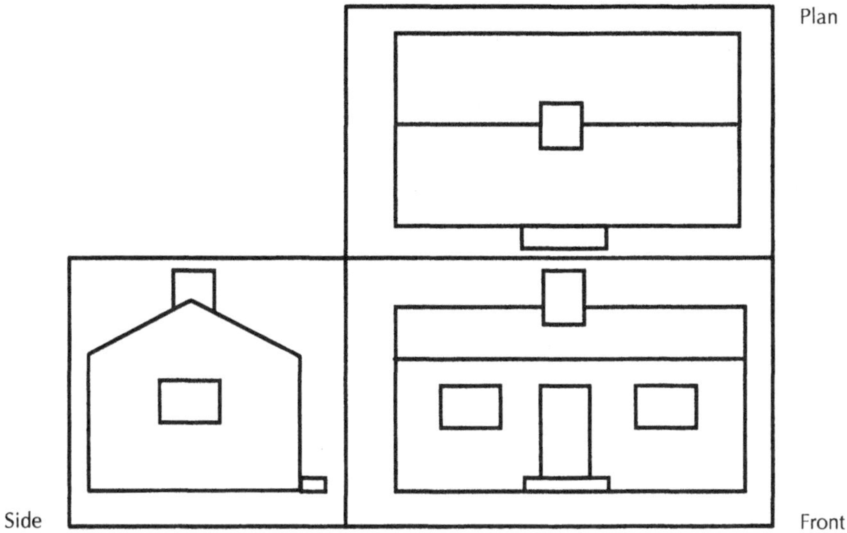

Figure 4.11

conventionally called "true" views. In the case in which any feature of an object lies at an angle to the standard reference planes—for example, the roof in the house in previous figures—it will appear foreshortened. Additional projections are necessary to show its "true" configuration and relation to the rest of the structure. If there were some design to be incorporated in the roof (a skylight or a pattern of shingles, like the names one sees on barn roofs), its true location, configuration, and dimensions could not be seen in any standard view. To see a true view of such a feature requires a new frame of reference, which is constructed perpendicularly to an edge view of the feature in question, as shown in figure 4.12.

This kind of view is called an "auxiliary" view. Its oblique (or "rotated") frame of reference is located in reference to the missing axis of the view from which it is derived: Here, the axis perpendicular to the line of sight is actually an edge view of an X plane. In such a view, notice that only *one* feature is shown in "true" view: the rear half of the roof, from which the view is oriented. Everything that is square to the world, except the overall length of the house, is foreshortened in this view. The perpendicular lines that define the local coordinates of this space are often called "work lines" but are, like the standard axes, technically the edge views of planes.

As far as the object is technically concerned, which frame of reference one begins with does not matter. Theoretically, one could begin with the auxiliary view, and construct the standard views as rotated projections from it. It is only for convenience and efficiency that designers begin with the standard views,

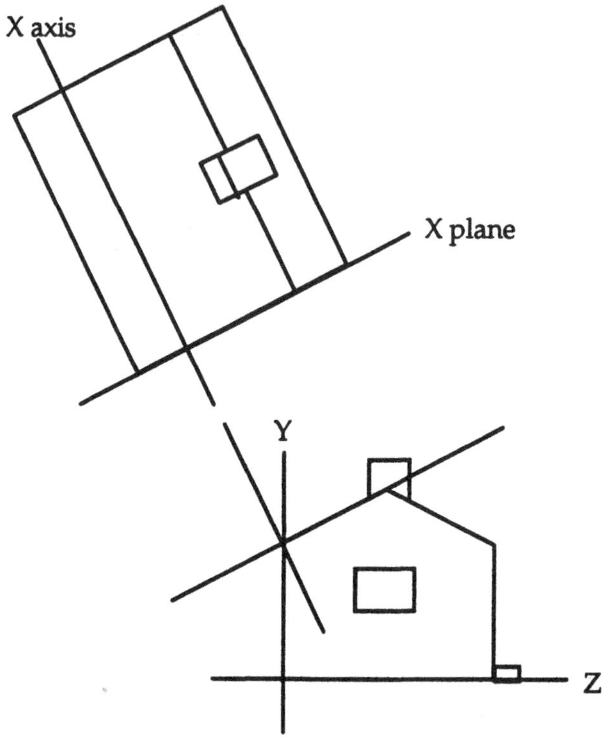

Figure 4.12

working from the general or most inclusive views to the specific and more detailed views. These views are also generally oriented so as to coincide with parallel and perpendicular features of the subject matter, which makes drawing easier.

One of the elegant aspects of descriptive geometry is the way in which these oblique frames of reference are easily set up and clearly related to features documented in the standard views. The experienced reader of such drawings becomes accustomed to seeing views from any direction, in any orientation; there is no absolute point of view, or any necessary up, down, right, and left in an orthographic drawing. It is the interlocking planes of reference, tied to the intrinsic geometry of the structure under consideration, that provide the overall understanding of the object or relation depicted. By measuring the angles of auxiliary views (based on the line of sight, or angle of projection) from the arbitrarily defined X, Y, and Z axes, it is possible to locate any auxiliary view accurately in respect to the standard views. One could just as well reverse this procedure, in which case the X, Y, and Z axes would be located in the auxiliary view; then, the standard views would be rotations.

One who only "reads" such a drawing must necessarily lack the intimate knowledge of it that derives from making the drawing. A designer who has constructed not only the lines but defined the space they are in "knows" this space and the object more thoroughly than anyone without that experience. Because all design drawing is done before the actual object exists, the drawing is more than a model or representation of the object. It is also a projection and embodiment of the thought process involved in arriving at the representation in the first place. After setting up an auxiliary view based on information from a standard view, one works out the oblique feature in the auxiliary view, and then projects these lines back into the other views where it will be shown in a foreshortened or rotated position. To design successfully, one must be proficient in translating between and integrating in the mind's eye all of these representations and frames of reference, regardless of orientation on the page, to arrive at an understanding of the "whole" object. This can be a difficult process.

Other systems of projection have been developed to arrive at more easily understandable, "pictorial" views. The two most widely used systems are *perspective* and *isometric*. Perspective is the most well known and actually predates systematic orthographic projection, but it introduces distortions of distance and angle (figure 4.13A). Isometric preserves parallelism and relative size, but it distorts perpendicularity (figure 4.13B). In both perspective and isometric views, however, all three axes of an object can be shown at once, and the viewer can understand what the object "ought to look like" much more easily.

To the eye trained in perspective, figure 4.13A has pictorial advantage; this is what the object "ought to look like." However, the edges of planes that would be parallel in the built object are not parallel in the projection, and the relative distances along the X, Y, and Z axes are distorted. Despite its pictorial

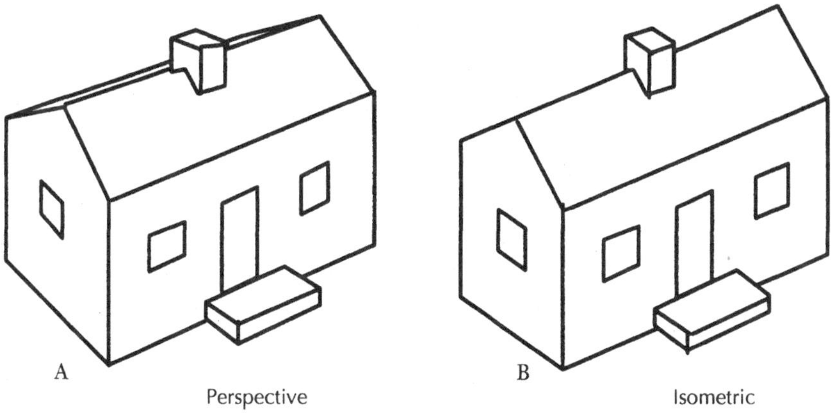

Figure 4.13

advantages, perspective involves elaborate systems of projection, and precise perspectives are much more difficult to construct than isometric or orthographic views. Consequently, perspective projections are not commonly used in engineering drawings. Its persistent dominance as a pictorial practice and mode of thought, however, is discussed at length in the next section of this chapter.

In isometric projection, the X, Y, and Z axes are drawn as equally inclined to the drawing surface, and equal distances along those axes are drawn as equal. Thus the edge view of the planes of the house are drawn as parallel lines, and the length of the lines in the X, Y, and Z planes are drawn in proportion to one another. To the eye trained in perspective, however, isometric projection looks curiously blocky and even introduces an illusion of divergent parallels: The far end of an isometric usually tends to look larger than the near end. The right, far side of the house actually looks larger than the left, near side. Isometric projection has advantages, however. Isometric views can be easily constructed or derived from standard orthographic views. They can be developed either through a series of auxiliary projections or, somewhat more easily, through what Booker (1979) calls a "secondary" system such as isometric drawing, in which the drawing is laid out along parallel axes 30 degrees from horizontal. In addition to the ease of constructing isometric projections, accurate measurements can be taken from isometric views. For this reason, they are probably the most common form of engineering illustration.

The power of descriptive geometry lies in this versatility: In a completed drawing, any one view can be derived from a combination of other views, and additional views can be generated as needed from what already exists. At the same time that it rationalizes the inconsistent practices of many empirical graphic disciplines, it offers the means to visually document forms in space of any shape or configuration. In the case of freeform curves and surfaces, this may be difficult or cumbersome, but it is at least possible.

Although curved surfaces can also be described exactly through mathematical equations (for example, computer graphics are actually a translation of a mathematical database of X, Y, and Z coordinates into pixels on the screen display), it is their graphic expression that is more fruitful for the human imagination. By developing or viewing a series of two-dimensional sections and putting these together with projections, a skilled reader of drawings can efficiently and accurately envision the shape that will result in three dimensions; this result is not as easily obtainable by reading its mathematical expression as an equation or a matrix of coordinates. Although mathematics may be useful for calculation—such as determining interior volumes or aerodynamic efficiency—numbers are not especially useful in assessing other spatial qualities of the shape such as the "fairness" of a curve (being free from bumps or flat spots) or its relation to other areas or features of a complex shape (notice how the A-pillars—the structure supporting the roof at the edges of the windshield—of an attractively styled car nearly always point to the center of the front wheels). These features can only be accurately developed through graphic projection and evaluated through the faculty of vision. Although the designer routinely uses words to describe the qualities of design features and numbers to quantify angles and distances, words and numbers cannot duplicate the imaginative and synthetic role of orthographic projection in identifying and solving problems of space. Descriptive geometry, as Comte observed (in the quotation that opens this chapter), is indeed applicable to "any figures whatsoever."

Drawing Things Together

For centuries, the geometrical art and science of perspective has served as both a set of graphic techniques and a model for thought (Damisch, 1994). But its insistence on a fixed and unitary point of view fails when confronted with the dynamic conditions and multiple viewpoints of our contemporary world, in which challenging individual and cultural possibilities—new dimensions, modes, and spaces of communication—are continually opening up through technological innovation and cultural change. Our recognition of these possibilities is limited by our attempts to put things "in perspective," that is, by understanding the task as a matter of seeing from one point of view or another. After five centuries of dependence on perspective—as a practical technique, a model for thought or, as Mitchell (1994) puts it, a figure for ideology—we may have reached the limits of what it can do for us. Perspective, as it turns out, is part of the problem, rather than part of the solution.

The entrenched power of graphic perspective, in combination with its textual equivalents—as demonstrated in the "equitone" prose of the modern novel (McLuhan, 1962) and the centrality of the scientific observer in technical writing (Bastide, 1990)—make perspective a formidable obstacle to the development of other models for thought. Even though the perspective-based practices of literacy and numeracy that emerged in the print era made modern science possible, we now know, thanks to Heisenberg's uncertainty principle (see Gregory, 1988), that the isolation achieved through perspective in the laboratory, the library, or the studio is but an intellectual fiction; it is not possible to remove the observer entirely from the events observed. "Seeing" is never a neutral activity. Perspectival seeing is not a neutral activity, either. What you see is not only dependent on your physical and cultural frame of reference, where you stand, and in which direction you look; the very act of seeing can change what is seen, and where and how you choose to look in the first place depends on what you want or expect to see. What one wants or expects to see is in turn shaped by structures or models of knowledge such as perspective.

Perspective, whether graphic, scientific, historical, or cultural, cannot readily account for a multiplicity of perspectives. The example of graphic perspective illustrates the difficulty. An accurately rendered graphic perspective appears—ironically—to the experienced "reader" as visually explicit and optically correct; perspectival distortions (such as the convergence of parallels and variance in size of objects according to distance) are generally read as accurate represenlions of the scene. However, the same scene, seen from a different graphic perspective, can read so differently that it can be difficult to recognize that the two—or several—different points of view actually contain the same subject matter. This is because any perspective, however detailed, can be no more than a partial and selective representation of a particular spatial configuration. And what a particular perspective does not show is as important as what it does show. Perhaps more important, perspective isolates the observer from observed. The artist (as viewer) is always at a specified distance and angle from the subject matter, as is the viewer of a picture (although these are not always identical).[19] Both views (of artist and viewer) are disconnected from the dynamic and multidimensional environment of which the picture represents only a single point of view at a frozen moment in time.

A systematic concept of how this happens can be developed by examining a well-known graphic example: the Dürer etching of an artist making a perspective drawing (figure 4.14). This image has been so widely reproduced as to constitute a visual trope for the work performed by perspective.[20]

In this picture, the artist's eye is stationed at the pointed end of a vertical post, which aids him in constructing a perspective drawing of a draped nude as seen through a wire-gridded panel positioned midway between them. Of course, the picture we are shown is completely different from the picture the portrayed artist is making. The picture *we* see was constructed from *another* point of view, outside the scene being shown. The angle from which we observe is nearly perpendicular to the line of sight through which the portrayed artist is viewing his own scene under construction. To show the "whole" scene as it "really" was—that is, Dürer in the act of drawing the artist at work—would require establishing yet another point of view in a third dimension, say, perpendicularly from above, which would be a "plan" view (from the ceiling). To show this scene would require projection to a different plane (the floor, perhaps) and therefore would produce yet another different perspective (and picture).

No one perspective, or point of view, in this case or any other, is sufficient to incorporate the entire scene. Even though each individual perspective

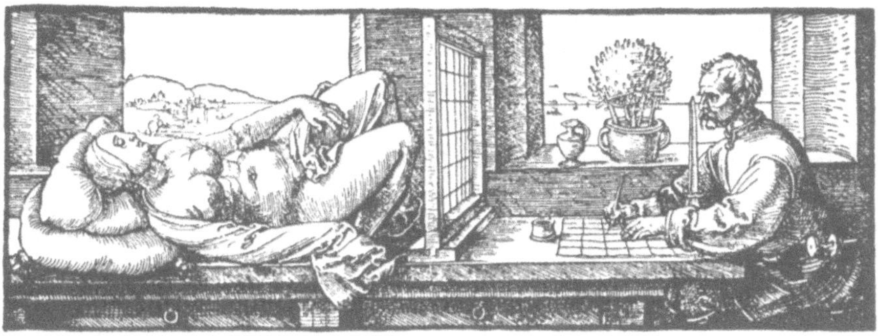

Figure 4.14

"contains" the others to a certain extent, each reveals details and relationships that are not necessarily available from any other. The differences in horizons and vanishing points among them make it difficult if not impossible to reconcile all of them and the different information they contain as being instances of a more complex whole.

Such limitations tend to be overlooked because of the apparent optical accuracy of perspective, especially in comparison with earlier and less rigorous techniques of projection. However, the power of perspective derives only partly from its illusion of truth. Perspective is also a circular argument: It shapes our vision, our depictions of space, and the space(s) in which we exist. It appears to prove itself. But perspective is only valid within the cultural context that has been built around it. As Snyder (1980) explains, perspective is the cultural "attempt to secure a scientific basis for picture construction that serves, in turn, to warrant the viewer's belief in the fidelity of the picture to what it represents. . . . The primary condition for this kind of picture making is the belief that vision is amenable to depiction because it is itself pictorial" (234).

Science appears to "prove" perspective by comparing it with the images achieved by photography. But this "proof" depends on a belief in the accuracy and precision of photography as the "correct" way of seeing and visually representing the world. It does not acknowledge that photography, as a product of mechanical engineering as well as chemical and optical sciences, was developed to achieve perspectival seeing. Nor does it acknowledge that many optical distortions are generated through strict adherence to the rule of perspective, as demonstrated by the pinhole camera, where, for example, spheres become ellipses (Pedoe, 1976). However, as new electronic technologies emerge enabling the essentially undetectable manipulation and distortion of photographic images (except insofar as such visual tricks defy common sense and the reality of experience), our confidence in the veracity of science, to say nothing of pictures and our own visual perception, can only be undermined. Despite our lack of skepticism when it comes to perspectival seeing, seeing, filtered through the apparatus of picture-making, broadcasting, and publishing technologies, cannot be "believed" any more than can hearing the speeches of politicians or reading the statistics of economists. It is necessary to know how information is gathered and assembled in order to evaluate the projections made from it. Perspective, in all its forms, makes such an evaluation a difficult if not impossible task.

Relying on perspective as a model for (parabolic) thought and projection is ultimately counterproductive because the perspectival insistence on a singular, dominant, and unitary point of view tends to obstruct the imagination of other possible (spatial) solutions. If we understand words, numbers, and graphics to provide different perspectives, it is possible to extend the above insights to say that relying on the singular perspective of either words, numbers, or graphics, taken alone, cannot account for multimodal parabolic relations. While language is particularly unwieldy in dealing with problems of space, numbers and pictures alone will not do the job either. It is far more productive to understand that words, numbers, and pictures are actually collaborators in producing meaning. To better understand this collaboration, I again turn to the way that problems of space (whether they appear to be primarily textual, mathematical, literary, or scientific in character) are both identified and solved through parable, in service of which nonperspectival graphic technologies (orthographic drawing, especially) are much more useful than perspective. To explore this potential, we need to extend the understanding of the process of parabolic projection beyond the study of parable as primarily a literary phenomenon.

Parable is deceptively easy to study through language, because it is visible and readily constructed in language; it is a concept that was developed to explain language. However, if parable is truly the basis of human thought it must be at work in other modes of projection as well. To investigate modes of projection beyond (but inclusive of) language, it is helpful to move outside language as the path to understanding parable. In fact, the process of projection is actually made more explicit and visible in graphic inscriptions (such as orthographic drawings) than it is in either textual or mathematical inscription. By studying the making of pictures, more so than the study of pictures themselves, we may arrive at a formal and scientific yet dynamic understanding of the nature and workings of projection. This understanding, in turn, can be projected to a better understanding of parable in other modes of projection.

In graphic or textual representation, projections are conventionally reduced to the two-dimensional space, or plane, of paper: the page. To think geometrically, we would systematically relate these different spaces or planes of experience and knowledge to one another. This would entail understanding their parabolic relation to both their subject matter and the construction of knowledge in general. In so doing, we might be able to comprehend better how different knowledges (such as geometry, history, and physics) relate to (project onto) one another from a more complex dimensionality, much as the Author of *Flatland* was able to step outside his own space and see it from a third dimension.

This move directly parallels Galileo's move in arriving at his laws of physics. Latour (1990) asserts that a "simple change in geometry used by Galileo [allowed] him to connect many different problems, whereas his two predecessors worked on disconnected shapes over which they had no control" (48). Latour never really explains the nature of this "simple" change in Galileo's geometry, but this change was anything but simple except in retrospect. (A historical point of view is not reversible; no one has 20-20 foresight!) Despite contradictory sensory and scriptural evidence—the unbroken belief in which apparently misled Jordan, Stevin and others—Galileo, "in the last instance, believed *more* in the triangular diagram for calculating the law of falling bodies than in any *other* vision of falling bodies" (51). Because he was able to place full faith in mathematical and graphical inscriptions, he was able to think imaginatively in ways impossible for others still bound by their belief in other (primarily textual and scriptural) forms of inscription.

In a quite formal sense, Galileo was a model geometrical thinker. He was able to free up his thought from the ambiguous and limiting dimensions and horizons of previous concepts (and representations) of space, and to relocate the problem into a multimodal frame of reference to draw clear correspondences between the "facts and the imagined reality"—between ideal and real space, and between events in space and graphic representations of these events. The orthographic Cartesian axes of coordinate space become here analogues (parables) for other dimensions and planes of experience—geometry and physics—that cannot be reduced completely to graphic (or to verbal or mathematical) representation. But they can be held conceptually in the mind's eye with the aid of graphics. "[I]n Galileo's mind . . . what gets connected are three different horizons held synoptically because the surface of the paper is considered as geometrical space" (48).

Even though geometry and physics are clearly two of the three "horizons" Latour refers to, he is never quite explicit as to what the third one is. Most likely he is referring to the representational plane of a graphic inscription on paper. It is here where the mental process of invention becomes visible. Like thought itself, "physics" and "geometry" are abstractions, that is, parabolic constructions. Even though we may comprehend the abstractions of a physicist's or geometer's imagination once they are manifest in text, equations, or graphics, these imaginary spatial relations are not directly observable. Neither is their primary subject matter. For example, physical materials or substances such as rocks and water, processes like fire, or even the spatial relations between objects in space are not observable in the text, equations, or graphs that project them. But the work of the imagination and the projections that are both the basis and the result of disciplined and imaginative thought must be *made visible* to be useful. This does not mean that a picture has to "look like" anything in the real world; its subject matter may be abstract, with no concrete equivalent. For example, Latour observes, a drawing showing the triangulation of forces is visible (and effective) in a way that the forces themselves are not. For it to "work" one must believe in the projection as evidence more strongly than one believes in other kinds of evidence. However, the evidentiary status of graphic projections is not especially amenable to either historical or scientific investigation with the tools we now have. Language and mathematics serve as poorly in the study of pictures as they do in the study of the geometrical and historical imagination. What we need is a set of tools better suited to the essentially spatial character of the imagination. Words and numbers may be necessary in developing a better notion of the imagination, but they are not sufficient by themselves.

Imagination, as the ability to visualize, is usually conceived as an internal process, something that happens in the mind's eye—a mental construction of a thing imagined or remembered. However, picturing to ourselves is a process inaccessible to observation. Since the study of communication is usually based on observable and measurable behavior and its products, one must focus on a specific kind of visible evidence of the imagination: inscriptions, which include pictures, writings, calculations, and the practices involved in the making and use of inscriptions as evidence. Descriptive geometry provides a basis for developing the formal ability to make precise and accurate graphic inscriptions, to make connections between these inscriptions and other kinds of inscriptions, and to demonstrate these connections by making further inscriptions. For these reasons, descriptive geometry has been of central importance to the industrial, commercial, and technological advances that have occurred in Western technological culture over the last 200 years. Thus descriptive geometry has been complicit in the production and reproduction of contemporary abstract space. But it also has the potential to provide a theoretical basis for the production of differential space.

How people make inscriptions and what they do with them demonstrates how people envision the world they live in: their past, present, and future. Graphic inscriptions—pictures in general, and technical drawings in particular—have unique and powerful communicative potential that has yet to be fully explained or exploited by investigations that reduce graphics into verbal or mathematical form. To develop their potential requires understanding how and why drawing and drawings are different from other modes of communication (words and numbers in particular), as well as understanding how drawing(s) function in relation to modes of communication as tools for the geometrical, historical, and parabolic imagination.

One key to this understanding is the concept of drawing as projection. Another key is to extend the concept of "picture" beyond the two-dimensional representation of things, or objects in space. Pictures are also parabolic projections of multidimensional problems and relations that are not always explicitly visible or recognizable but are nonetheless available to a proficient reader of drawings who may reconstruct or infer such information from the evidence at hand.

Among artists and designers—specialists in the generation of graphic inscriptions—the imagination is formalized and expressed through various practices that result in the translation of a mental construction of the already seen or as yet unseen into still another inscription: a graphic "text" visible by others. These graphic texts, like mathematical or verbal texts, are likewise reductions of complex and multidimensional phenomena to a medium of less complex dimensionality. Typically this involves reduction to the two dimensions of the page or, increasingly, the two dimensions of an electronic display. Seen in isolation from one another, or seen only as mimetic representations, inscriptions of any kind can provide only trace information. However, by taking several inscriptions together, in what Latour calls a "binocular" view, one can imaginatively construct multidimensional knowledge that is explicitly unavailable in any single inscription.

This is a specialized and practical kind of geometrical thinking, which is demonstrated routinely among architects, engineers, and others who read blueprints, a genre of technical drawing not usually conceived as particularly imaginative or inspirational.[21] Technical drawings are generally meant to constrain the range of actions available to those who use them. They not only inscribe and describe, they prescribe. The typical objective behind making and using such drawings is to arrive at the accurate and precise production of an object or assembly that more often than not must fit and work with other objects and assemblies. Ignoring or violating the instructions contained in a drawing generally results in undesirable consequences. These can be relatively minor, as when machines wear out prematurely. But they can also entail catastrophic failure: buildings collapse, airplanes explode in midair, ships break up in heavy seas.

The discipline required to produce and read technical drawings in some ways inhibits, and is perhaps inimical to, the exercise of the imagination. Seen in isolation from other inscriptional practices, technical (orthographic) drawing can easily be seen as a limited and limiting endeavor. However, these inhibitions and limitations are neither necessary nor inevitable. Indeed I contend that imaginative thought generally is based in the disciplined ability to visualize, a skill integral to the production of all kinds of drawing. Consequently the disciplined ability to visualize ought to be more widely cultivated through drawing instruction and practice. But further, this ability should be extended by developing deeper knowledge and understanding of the formal methods of geometrical thinking that are associated with the production of technical and scientific graphics. Even though all (parabolic) inscriptions—textual, graphical, or mathematical—are spatial in character, technical (orthographic) drawing is perhaps the most explicit, accessible, and well-developed language of space that we have; it ought to be widely taught and used to understand and solve problems of space.

The theory that subsumes orthographic drawing practices is based in descriptive geometry. Orthographic drawing—much of which is rooted in empirically developed and inconsistent theory—when informed by descriptive geometry becomes more than a set of graphic techniques for the representation of objects and problems in space. It is, like its cousin perspective, a model for thought. By its "positive acceptation" (again referring to the quotation by Comte that opens this chapter[22]) one is drawn into the production and representation of new social space that is uniformly and accurately observable, measurable, and representable. Unlike perspectival space, which always distorts dimensions and other spatial relations no matter what point of view is chosen, the space of descriptive geometry insists on the representation of "true" distances, angles, and other spatial relations. Furthermore, the practice of descriptive geometry, which requires the mutual involvement of hand, eye, and mind in drawing, reconfirms the muscular and tactile character of Euclidean geometry (Ivins, 1946) and "absolute"space (Lefebvre, 1991). Thus it works against the compartmentalization of theory and against the distortion of practices and representations of this space.

Descriptive geometry makes space (abstract space, in particular) available in a way different from the way perspective does. Perspective, as either a model for thought or a technique of representation, is an inadequate means to represent the dynamic and multidimensional character of any mode of social space: absolute, abstract, or differential. Objects and relationships will always look different depending on the point of view taken, but it is difficult within a perspectival framework to reconcile these differences. Perspective, put into service as an instrument of cultural power, tends to deny the validity of "alternative" views, as it unifies space by reducing or subordinating all points of view, dimensions of space, and its subject matter to a single dominant one: the two dimensions of the surface on which inscriptions are made.

To a certain extent, we automatically restore some of what is lost in this reductive process when we read a perspective inscription as an image of a three-dimensional scene. Depth, which in common drawing parlance is called the "missing" axis, is not representable on the two dimensions of the page; it is implied through perspectival conventions such as foreshortening and the convergence of parallels, the knowledge of which tells us how to construct a mental representation of the scene. However, any perspective, no matter how detailed, can be more than a partial representation of the space in which it occurs. Here, I mean "partial" in two senses. First, a perspective is a fragmentary view. Second, it is always based in certain interests. What it does not show is as important (from some other perspective or from some other interest) as what it does show; and what is left out cannot be readily reconstructed.

Orthographic projection, in constrast, insists on a multiplicity of views and provides a unified perceptual, logical, and mathematical framework (through descriptive geometry) that enables the integration of explicitly partial and limited representations—views—of both imaginary and real objects and problems within an intelligible and systematic multidimensional space. Information not explicitly contained in a completed drawing can readily be inferred or reconstructed by anyone who knows the practice and language of drawing.

The fact that so much drawing practice is rooted in empirical rather than theoretical knowledges is but one obstacle to its application to problems of space. The established dominance of literature and mathematics as the preferred languages in contemporary technological culture is probably a far more formidable obstacle to any such application. Resistence to the development and acceptance of multidimensional space can be explained in part by the (partial) interests that have historically been served by language-based and logicomathematical modes of thought. At the same time that the Renaissance made possible a liberation from the "trammels" of Classical thought and knowledge (Ivins, 1953), such inventions as the printing press and movable type actually served to reinforce the historical priority of language-based and logicomathematical modes of thought that were first demonstrated in Plato's *Republic*. Put quite simply, with the relatively sudden and widespread availability of cheaply produced texts, the project of attaining widespread literacy became an urgent priority at all levels of culture. Thus, the culturally privileged place given these modes of thought, rooted in a Classical mentality and transformed in the print era, has served for centuries to obscure the potential of drawing as projection in the development of the imagination.

Eisenstein (1979) describes at length the crucial role of printed texts in the advances made in scientific knowledge by such key figures as Brahe, Kepler, Copernicus, and Galileo; however, the concomitant effect of printed text in empirical disciplines, such as the publication of heretofore secret trade knowledge, has been less recognized. Ivins (1953) cites the invention of the "exactly repeatable pictorial statement" as a key element in escaping "the trammels of Greek ideas" (23), which had served for centuries as a roadblock to the advancement of thought. Such pictorial statements may also be a key element in the success of print-based knowledge, especially that of a technical and commercial nature, in Western European culture. The invention of perspective and the development of more sophisticated tools and methods for other kinds of drawing are closely related to the rapid dominance of text as a mode of communication and are of perhaps equal importance in the development of modern science and technology. As a result, the conceptual importance of drawing as projection that perspective made possible was quickly overshadowed by the role of print in the advancement of scientific knowledge. Its importance would not be recognized for nearly 200 years until Monge combined his theoretical background with his knowledge of several empirical disciplines to arrive at the invention of descriptive geometry.

For the past 200 years, both the empirical practices and theory of descriptive geometry have been widely published and accessible. But neither has much informed the spatial experience of the general public. Without specialized knowledge and training on the part of the "reader," texts on the subject, the graphic practices they demonstrate, and the documentary evidence produced through the exercise of these practices (drawings themselves) remain largely unintelligible to the world at large. The lack of technical visualizing skills among the general public is well known to the practitioners of design disciplines, such as architects, who must often try to explain what they have conceived to an audience that lacks their visual language. Even the best of perspective renderings and models are inadequate to communicate what objects—from public buildings to home remodeling projects—will "really" look like. If people cannot even read a graphic representation of a concrete thing such as a building or an automobile, it is too much to expect them to read or understand a graphic representation of an abstract problem that may require the integration of multiple views to arrive at a comprehension of the whole.

This widespread lack of formal visual and spatial knowledge is increasingly a disadvantage. Today, Davies (1990:16) observes, we live in the midst of an "explosion of pictures." However, most of these pictures don't really qualify as technical graphics, even though their representational content may be technical, in the sense that they depict the products of technology. This highly visual environment shapes our thought just as powerfully as does the experience of literacy; we have come to expect pictures as enhancements to text. To the reader (or viewer) accustomed to multiple media, a verbal message unaccompanied by some kind of pictures seems somehow incomplete, unduly difficult perhaps, and even worse, uninteresting. In striking contrast to our reliance on the visual, few people have the requisite skills to produce it. Isn't it odd in this visual environment how many people claim (exaggerating perhaps only slightly) that they "can't even draw a straight line"? And without even the most basic skills needed to produce visual messages, how can we expect people to be competent "readers" of them?

What Mitchell (1994) and others call the "pictorial turn" of contemporary culture is an implicit demand for a better means of dealing with pictures. As Davies implies, we ought to learn to "write" pictures as well as we "read" them, although my contention is that we don't read them well either. Even though Davies is trying to suggest the possibility of a visual analogue to verbal literacy, his choice of metaphor is unfortunate (as is my own unhappy reliance on "read"). As Mitchell observes, pictures are not necessarily explicable on the basis of a literate model (such as Messaris, 1993, seems to suggest). Lacking the widespread knowledge of a formal discipline of the visual sense, such as descriptive geometry and orthographic drawing provide (though in a specialized sense), we can hope for little more than to translate what we see into written or spoken words and to see pictures as little more than mimetic representations, and in the process never gain a sense of the fundamentally different content of, and work performed by, pictures as projections.

That we continue to insist on the differences between words, numbers, and pictures, rather than looking at the ways in which they work together in producing meaning and knowledge, is but one demonstration of the existence of deep cultural and disciplinary divisions (summarized by Latour in the label "the Great Divide") that are only now beginning to be thoroughly examined by philosophers, critics, and historians. These divisions are based in the simultaneously fragmenting and centralizing tendencies of "abstract" (perspectival) space, and the cultural privileging of textual and mathematical forms of knowledge and communication (Gardner, 1983/1993; Lefebvre, 1991). This simply must be addressed if we are to work through the problems of "abstract" and "contradictory" space toward an imaginative understanding and production of "differential" space.

I am not suggesting that we look to descriptive geometry as an answer, but as a point of departure. It is still an open question whether social space can, or even ought to, be geometrized along the lines of physical or mathematical space, as perspectival space has been for centuries. Furthermore, as long as the theory and practice of descriptive geometry is coupled to the concepts and interests that underpin and produce "abstract" space, this prospect is likely unattainable. A large part of my project here is to open up this relation through a study of the role of drawing theory and practice in the production of knowledge and in the solution of spatial problems (both abstract and concrete), rather than in the production of things.

Several hundred years ago, drawing was an important part of educational curricula, but visual education gradually has been "eclipsed" over the last several centuries (Stafford, 1994). This omission ought to be corrected, but substantial disciplinary and cultural obstacles stand in the way of any remedy to the situation. First is the historically consistent identification of drawing with art, craft, and other vocational (empirical) practices, and the general division of these practices from professional (scientific) practices that are based in writing and mathematics. Second is the stratification of society on the basis of one's situation in these various divisions (such as being the son of an itinerant peddler, or a Bonapartist, or an artist rather than a scientist). Third, a residual effect of the ancient Greek mentality, is the ongoing cultural trend toward increasing specialization and compartmentalization of knowledge based in the privileging of language and logicomathematical forms of communication, which is demonstrated in the established "modes" of script and print communication and which appears to persist in the emerging mode of electronic communication. These several reasons serve both to perpetuate the dominance of established forms of knowledge and systems of power and to limit the potential of drawing and other forms of visual communication to contribute to the development of a discipline of the spatial and historical—geometrical—imagination.

These factors can be seen conspicuously at work in systems of public and private education. The study of visual communication practices at all levels is typically isolated into areas such as fine arts, graphics, and illustration. At an elementary level, art programs are often little more than "luxuries" that are, along with music, the first subjects eliminated when budgets are cut. These areas have little overlap with the other liberal arts such as English, history, and so forth, which emphasize literary proficiency through such practices as writing, criticism, and textual research. In disciplines that are less explicitly verbal, such as science and engineering, an emphasis on mathematics dominates the curriculum.

In engineering, particularly, this is an unfortunate limitation. Even as recently as a few decades ago, drawing, like other forms of hands-on experience, such as working with machine tools, was a basic component of most engineering programs. Since then engineering educators have increasingly promoted the status of engineering as a science. Today, as a consequence, both teachers of engineering and the employers who hire newly minted engineers lament the lack of drawing and design ability among students and graduates (Booker, 1979; Ferguson, 1992). Even so, engineering programs continue to emphasize analytic methods, becoming ever more mathematical. Today, as other subjects have "continually whittled down the time that could be devoted to drawing" (Booker, 1979:152), most engineering students have little more than a basic knowledge of either descriptive geometry or technical drawing, compared with mathematics.[23] In mechanical engineering, a discipline dedicated to the solution of spatial problems, this is an unaffordable omission. Engineers ought to be, first and foremost, geometrical thinkers *par excellence*, true innovators in the art and science of imagination. However, without the appropriate conceptual tools and the physical skills to make their thoughts visible and comprehensible to a wider public, their potential as innovators is severely curtailed. If engineers cannot clearly or accurately envision and draw even the internal structure and function of a designed object, we can hardly expect them to envision or communicate its relation within a larger cultural context. Engineers focusing exclusively on the development or improvement of technologies, as so many do, limit their potential as innovators as they avoid cultivating the discipline of thinking geometrically.

Technology develops through the practice of design, and modern industry would be impossible without engineering drawings. Technological development is not culturally neutral, however, and technological progress is not inevitable (see Feenberg, 1991; Winner, 1986). Design of any kind, as Buchanan (1995) points out, is more than an aesthetic or technical endeavor; it is a potent form of environmental and cultural intervention. It is one (admittedly difficult) thing to generate new scientific knowledge and develop novel technologies, but it is an even more formidable task to envision and anticipate the wider consequences of such work. Science and technology change our lives, for better and worse, and it ought to be the responsibility of those who develop science and technology to be skilled at envisioning consequences and giving conscious shape to the future.

On the other side of Latour's "Great Divide," artists, like engineers, also generally fail to exploit the potential of drawing and graphic theory in the construction of knowledge or to examine the effects of their work on culture at large. Because in many ways art and technology are converging in the emerging electronic, and increasingly graphic, communication environment, the role of artists as cultural visionaries ought to be recognized and promoted. However, their innovative potential is currently diminished along several different but geometrically related dimensions: in the relationship between artists and their tools; in the nature of the workplace; in the persistence of the distinction between fine art and the practical arts; and in new disciplinary distinctions designating roles in the production process. Even though artists of many different kinds are the primary generators of the images that flood our daily lives, they are typically not involved in the invention, production, or maintenance of their own tools or materials. When artists are held apart from this kind of relationship with their tools and materials—whether in making their paint and brushes or designing their own computer program—they are limited in their ability to think with and through them. Their work tends toward the translation of words into pictures rather than what James Elkins (1999) calls the "alchemical" transformation of the very substance of the world. In addition to the divorce from an intimate bodily relationship with tools and materials, artists are often not involved in deciding the subject matter of their productions. Artists are increasingly dedicated to the generation of new products, whether a hot new machine or a snappy illustration or advertisement, to satisfy a competitive consumer market. This is especially true when artists are employed by a commercial enterprise, that is, when they "work for hire." And given the status of artists in contemporary culture, it is difficult to even be an artist if one does not "work for hire." The potential to envision art in its visionary role is further diminished by the old distinction between fine arts and practical arts. As the distinction itself conveys, there are practical arts such as illustration, drafting, and engineering, and there are fine arts, which are *not* practical. But what could be more practical than envisioning and transforming possibilities for the way we live? The artist as visionary is further fragmented by new forms of disciplinary distinctions that overlay and transform the distinction between fine arts and practical arts. Increasingly, social space is shaped around the designation of the roles of artist, designer, user, or operator. It matters how someone who works at a computer graphics terminal is defined. Is this person an artist? A designer? A user? Or just a machine operator?

The projection of these dimensions onto one another reproduces the simultaneously fragmenting and homogenizing tendencies of abstract space. For example, graphic manipulation software such as painting programs and CAD (computer-aided design) programs are each designed with completely different capabilities and performance characteristics. However, each computer user is expected to conform to the different interfaces generated by the differently designed software. But there is something further at stake: Users must conform to interfaces conceived along the lines of a cultural division of artistic labor: fine art artists (painting software) versus illustrators (CAD). And both designers and users must conform to expectations, possibilities, and demands conceived along other dimensions of the cultural division of artistic labor: designer (producer) versus user (consumer). But designers too are affected by these distinctions, as the design process is itself fragmented as artists of all distinctions become isolated from the production and maintenance (and in some cases, use) of their tools and materials. The design of computer hardware and software is the province of engineers and programmers, who more often than not have no acquaintance with the actual practices and needs of artists and designers. To envision the complex connections among these dimensions, to notice how they "spin out" of two-dimensional space and hold them in the mind's eye is a significant challenge; to envision ways to transform these relations is more challenging. We can only begin to appreciate and work with the complexities by thinking geometrically. In the meantime, this construction of abstract space—its problems and contradictions—works to keep artists, like engineers, from thinking and living geometrically.

It might seem that computer graphics—as a form of drawing—could work to overcome the fragmentation of the designer and user. But computer graphics both reinforce and undermine the notion of drawings as projections. At the same time that the glowing display generated electronically is directed by the mathematical coordinates of an object in the database of computer memory, the physical and conceptual involvement of the artist in the drawing process is transformed and diminished. The kinesthetic experience of manual drawing—either freehand or with tools like compasses and triangles—is an essential component of both getting to "know" firsthand the space of the object being drawn and developing the ability to think spatially and visually in this space, which is in turn necessary to the production of consistent and systematic images. This experience is nearly completely lost in computer graphics. The physical skill and hand-eye coordination of an accomplished artist is reduced and replaced by coping with the clumsy movements of a mouse, the disconnected movements of a stylus on an electronic sketchpad, or the typing of commands and coordinates on a keyboard. All of these feats of constrained dexterity force one out of a visual mode into a verbal or mathematical mode of thought. Even though the ability to construct and transform complex three-dimensional objects is in some ways greatly facilitated in the electronic space of the computer, the technological disjunction of artist or designer from direct involvement in the space of his or her own work, and the incorporation of the (now) user into the uniform and Euclidean character of virtual space, again demonstrates the simultaneously fragmenting and homogenizing tendencies of abstract, and perspectival, space.

If any inscription—knowledge made visible—is to be effective, it must be intelligible. The investigation of drawing and descriptive geometry theory can serve to develop this intelligibility; computer graphics technologies serve to obstruct it. As long as we lack any widespread knowledge and practice of a true discipline of vision, the geometrical and historical imagination will continue to be constrained by the priority of language and logicomathematical thought. The hope of either a "science of space" or a "quasi mechanics of history" will remain unrealized, and the (parabolic) information "contained" yet not made explicit in pictures, like the theoretical possibilities opened up through an investigation of this information or knowledge, will remain unintelligible and inaccessible.

Technologies of electronic communication are changing the way we think about, and act in, space. We are in the midst of a transitional era, in which the magnitude of cultural change will no doubt be as great as it was in the era of the ancient Greeks, when the invention of the phonetic alphabet enabled script and literacy, or in the era of the Renaissance, when the invention of movable type and the printing press made accurate texts and pictures cheap and accessible for the first time in history. These changes parallel the production of, and transitions between, the different modes of absolute, abstract, and, now, differential space. Literacy, a model for thought enabled by the invention of script, made possible the transition from absolute to abstract space. Perspective, as a model for thought, has served to extend abstract space as well as the dominance of Western technological culture to encompass the entire globe. Descriptive geometry can serve as the basis for developing and projecting nonperspectival models for thought more appropriate to the mode of multidimensional, differential space.

Even though something was lost in each of the transitions between modes of space, something was also gained. Being literate, we no longer experience poetry as oral cultures know it; and we have lost the memory skills essential to the mode of orality (Ong, 1982; Yates, 1966). At the same time, literacy contributed to the development of logic, philosophy, and science. The question of what might be lost in the transition to differential space is difficult to answer. In trying to think geometrically, I prefer to think of what we might gain: most of all, an effective means of dealing with the accumulated contradictions of abstract (social) space and the multiplicity and plurality of global technological culture. These cannot be dealt with effectively through the model of perspective. Rather than constantly seeking new solutions or fixes, especially technological ones, to the problems of abstract space, I believe that it is time to envision new models for culture and to make the transition to differential space.

As I have argued, descriptive geometry can play a significant role in this transition. As a model for thought, descriptive geometry provides a basis for both integrating and preserving the integrity of multiple views and multiple frames of reference. As a systematic and empirically rooted model for parabolic thought, it is basic to the formal development of a structured, coherent, and multidimensional imagination in many engineering disciplines and can serve as a model for imaginative thought in other contexts as well. Descriptive geometry provides a theoretical basis for developing a multidimensional model of the parabolic relation of words, numbers, and pictures through the practice of orthographic drawing. Through cultivating the promising yet unruly potential of the geometrical imagination and utilizing the power of descriptive geometry as a model for thought, we may learn to make more consistent and coherent projections of ourselves, our world, and worlds to come, and translate these projections into reality. Inscribed knowledge—words, numbers, and pictures—are all parabolic projections onto different planes of limited dimensionality. Rather than assert the necessity or dominance of any of the singular points of view of language, mathematics, or graphics (a perspectival exercise rooted in the mode of abstract space), we need to find ways of drawing them together. The multimodal discipline of design may serve as the basis for developing an understanding of how inscriptions produce social space. Space is not something "out there"; it doesn't just "happen." Space is directed and intended, in a word, *designed*, through human activity.

CHAPTER 5

DESIGNING SPACE

Both "design" and "space," like spoken and written language and mathematics, are learned through firsthand experience. As I argued in the previous chapter, this knowledge can best be gained through the experience of spatial (that is, graphic) practice. But this is a book, and I am reduced here to using words as the primary tool to provide several partial experiences of design, space, and the relations between them. Through the exercise of geometrical thinking, their limited horizons can then be projected onto one another to arrive at a richer picture, or conception, of the significance of designing space. I envision this project much like Galileo's combining the horizons of physics and geometry, but the conception of space I reach for is social, not just mathematical. This picture, whether formulated in words, numbers, or pictures, is in itself also necessarily incomplete. However, it is designed to be taken up parabolically and combined with other partial knowledges that each reader might project onto it. The collective result will be the production of a space of greater dimensionality, in short, an exercise in thinking geometrically. Because geometrical thinking is synthetic, dynamic, and multiple in character, it is a uniquely appropriate way of envisioning and arriving at solutions to cultural and intellectual problems that are not amenable to other approaches, but which may fruitfully be understood as problems of designing space.

This is a challenging task. Building a spatial knowledge through language alone may be an impossible task, which is why pictures were used in previous chapters. However, pictures themselves are also insufficient. Taking parable as the basic unit of knowledge common to all modes of thought may make the task manageable. The fundamental parabolic relation I wish to explore is that between *design* as a structured human activity and *space* as both the context and the product of this activity.

The phrase "designing space" embodies three distinct yet connected (parabolic) relations.[1] The first, and most obvious, is the *design of space*, which includes the various practices, processes, and objects that structure our cultural and physical spatial environment. Words, numbers, and pictures are the primary tools for accomplishing the design of space. Textual documents, pictures, and built objects are the concrete evidence of this activity. Second is the *space of design*, which refers to the many different concrete and abstract frames of reference in which the designer operates and through which the imagination is inspired and/or constrained. The space of design includes "real world," "studio," and theoretical frames of reference. Perspective constitutes the currently dominant frame of reference. Third, to indulge in a bit of wordplay that I hope will not obscure its importance, is the notion of *de-signing space*, which means giving up our cultural reliance on the ability to account for and solve problems of space exclusively through language-based and logicomathematically based methods. The analytical acts of labeling, categorizing, and measuring spatial and visual objects, relations, and events all have theoretical and practical importance, but these have yet to lead to an adequate explanation of the visible and invisible effects of design on our lives: how we conceive, act, and produce space, that is, how we synthetically design ourselves and our lives. In short, I am after a sense of how we "live design."

If space is a product of human effort and imagination that has a history, the next step in this project must be to examine the parabolic construction of this history through the making of inscriptions. This requires not only looking at the documentary evidence of words, numbers, and pictures, but understanding how the practices of both constructing and using them participates in the design of (social) space. This is an exercise in parabolic and geometrical thinking.

As demonstrated by Jason Bourne and the poor Author of Flatland, a socially useful discipline of geometrical thinking is likely to be antidisciplinary and antisocial. We would not expect it then to emerge from within the existing structures of either the sciences or the humanities, which have deeply vested interests in language-based and logicomathematically based (perspectival) thinking. In fact, the dominance of these modes presents a persistent obstacle to the systematic investigation of geometrical thinking. The conspicuous priority placed on the widespread attainment of literacy and numeracy is characterized by the near-universal pursuit of mastering the three R's: reading, 'riting and 'rithmetic. The acquisition of these skills, which has for centuries been central to the spread of Western technological culture throughout the world, has served to create and perpetuate an environment where other competing modes of thought are simply overshadowed. The faculty of vision—and of envisioning—is so thoroughly disciplined by the narrow requirements of literacy and numeracy that one who has mastered these skills is poorly equipped to imagine, much less put into practice, other ways of seeing and producing social space.

Design theory and practice is currently so fragmented across and submerged within multiple disciplines, among which design serves multiple purposes, priorities, and interests, that it is perhaps stretching things to consider "design" as a unique and distinctive discipline in its own right. Even in disciplines such as industrial and graphic design, in which design is a visibly central priority, little theoretical agreement exists as to what design is or ought to be.[2] However, the indeterminacy of both the subject matter and the field of design is actually here an advantage; design, as an emergent discipline, is not (yet) bound by the theoretical allegiances and obligations that often constrain the scope of other established disciplines.

Richard Buchanan (1995) proposes that the emerging field of "design studies" might bring theoretical unity to the diverse practices of design. Yet he argues that what distinguishes design from other humanistic (and scientific) disciplines is the characteristic *indeterminacy* of both its subject matter and its methods. Neither are necessarily given to the designer, whose first task is often to arrive at a suitable problem definition that both constrains and inspires the imagination. In arriving at a problem definition, the designer may be required to develop new methods of representing it in order to achieve a solution.

The results of recent work in design studies are as yet as indeterminate as their subject matter. For example, Buchanan defines design as a diverse unity of all the arts of making, an essentially human activity whose essence "lies in the fact that human beings determine what the subject matter, processes, and purposes of design shall be" (55). However, Buchanan characterizes design theory and practice as fundamentally rhetorical in the classical sense. He reduces it to "how products come to be as vehicles of argument and persuasion about the desirable qualities of private and public life" (26). In taking this position, Buchanan reduces the distinctive and central role of the spatial and geometrical imagination in design to a limited rhetorical dimension. Design certainly can be about the making of things and the processes of argument and persuasion that bring them into being, but it is neither necessarily these nor reducible to them. To design is to envision or invent new relations between existing objects and concepts, not just to make things. Indeed, making things is actually an activity distinct from, although connected to, the rhetorical aspects of design. As the act of making connections between the imagination and reality, design is, even more fundamentally, the construction of parable that projects the overlapping and interpenetrating roles of words, numbers, and pictures in the production of space. Design is ultimately a potent form of social action or, perhaps more accurately, cultural and environmental—spatial—intervention.

Graphics—pictures—are to design as writing is to history or literature, or numbers are to mathematics. Pictures, as documentary evidence of graphic activity, *are* somehow distinctively different than words and numbers, but this difference cannot be fully explained by or reduced to either.[3] If talking or writing about pictures could somehow lead to conclusive knowledge of what they are and how they work on us and for us, one would think that centuries of effort by many of our best thinkers would have already led us to such knowledge. It hasn't yet, and it probably never will. Mitchell (1994) argues that all approaches to explaining the power of pictures that are based on the priority of language-based knowledge are bound to fail. He suggests that, instead of theorizing pictures, we might "picture theory" (16). That is, we might look at how pictures represent and embody not just visual but literary and other forms of knowledge. Even this inversion of perspective, however, is limited in its theoretical utility (it is still a perspective). Pictures, after all, are static evidence. They tell us relatively little about how and why they were made, even though much of this may be reconstructed by a proficient "reader," especially when accompanied by other documentary evidence. If, along with Mitchell, we are to accept that words, numbers, and pictures are dynamic collaborators rather than competitors in producing meaning, we ought to be studying how this collaboration occurs, that is, studying the practices and processes involved in their production, rather than the objects (such as documentary evidence) that result from this production. The study or analysis of words, numbers, and pictures as static inscriptions—objects fixed in space, either together in isolation—can lead to only a partial understanding of their synthetic and dynamic role in producing knowledge and social space.

Bruno Latour (1990) argues that we ought to examine "the way in which groups of people argue with one another using paper, signs, prints and diagrams" (21) and to "consider in which situations we might expect changes in . . . writing and imaging procedures to make any difference at all in the way we argue, prove and believe" (23) in order to gain a fuller understanding of the interrelated role of words, numbers, and pictures in the construction of scientific knowledge. When Latour uses the term "inscription" to refer to all three, he intends to draw attention to their shared rhetorical character. Ultimately, however, he focuses much more on the concrete documentary evidence of inscriptions (a logical emphasis, since a paper record is all too often all we have for historical investigation) and their role in the cultural (and rhetorical) exercise of power than he does on how people actually design, make and use inscriptions on an everyday basis.

To draw attention to the larger work of inscriptions—as playing a role not only in the design of space, but in the space of design and in de-signing space—I propose instead to use the term "transformation." Transformation, unlike inscription, calls attention to the fact that all inscriptions—textual, mathematical, or graphical—project onto each other, as well as to their subject matter or content, and are parabolically equivalent in the production of meaning.

Transformations of Space

To understand the key role of graphics in the construction of knowledge and in the production of space it is necessary to understand the making of pictures as well as we understand the writing of verbal and mathematical texts. Just as an inability to write limits the skill with which texts are read, the inability to draw limits the skill with which pictures are "read." By internalizing the skills of graphic production on a level analogous to literacy and numeracy, one can gain a fuller comprehension of the multimodal context that is jointly constituted through words, numbers, and pictures.

The previous chapters have been dedicated to an account of the theory and practices underlying the geometrical (spatial) imagination and the production of (ortho)graphic inscriptions throughout the history of Western technological culture. This was done, not to assert the priority of graphics over language and mathematics, but to serve as foundational knowledge for establishing the underlying parabolic character of literary, mathematical, and graphical thought and expression as different dimensions or "modes"of space that cannot be grasped through its projection onto any single plane of thought. Thinking geometrically, as demonstrated in the formal practice of orthographic drawing, is a way of assembling partial and limited two-dimensional projections into a more complex multidimensional comprehension. Even though a complex and multidimensional "whole" may elude all efforts at representation in its entirety, a coherent and systematic approximation in three dimensions is an improvement over an approximation in two dimensions.

The longstanding distinctions and corresponding disciplinary divisions among textual, numeric, and graphic modes of thought and expression are becoming increasingly unsustainable in our contemporary era of computer-based communication. Today, words, numbers, and pictures interconnect in the new multimedia environment. It is not yet clear how these new modes of communication will transform us and our culture as well as traditional cultures as yet untouched by literacy and numeracy. It is clear that what Poster (1990) calls the "mode of information" is more than anything a mode of *transformation*. There is no such thing as "raw" information; data are from the start selected, filtered, and structured (put in formation), both through our own cognitive patterns and processes and the patterns and structures required by computer programs and databases. However, stored as binary digits or as X, Y, and Z coordinates in computer memory, a database has no particular formation. For example, three-dimensional computer-aided design data cannot be displayed directly on a two-dimensional screen. The viewer must first select some view direction, scale, and center of interest to arrive at a display. The visual organization imposed by these selections is temporary and can be changed at any time; there is no such thing as a single "true" view of complex objects. There may be a "best" view, one which is less ambiguous, or appears to be more realistic, or is more useful than others. But no one view is any more than a partial representation of the database, which cannot be expressed in its entirety through any projection into a medium of lesser dimensionality. To the computer, there is no difference between perspectival, orthographic, or any other kind of projection on the screen. As long as graphics software is appropriately programmed, all these views are available at the touch of a few buttons. However, the computer cannot help the designer choose which is the "best" view. This choice must emerge from the designer's imagination and the goals of the design task at hand, which are not always easy choices to make given the wide range of alternatives.

Thinking geometrically may be the best means we can develop to cope with the multiple, pluralistic, and decentered worlds emerging around us. If, as contemporary physics has it, physical space consists of 11 or more dimensions, no single representation of it even in three or four dimensions is going to be complete. Social space can be no less complex. We don't know what all the dimensions of social space might be, much less have an idea of how many dimensions it might have. Any "view" of either physical or social space, like any representation of a database, must be seen as associated with, or based in, the interests and concerns of whoever constructs that view. If we are restricted to representational techniques and a cognitive apparatus that can cope with at most three dimensions that it cannot avoid further reducing to the two-dimensional space of the page or the computer display, a space of 10 or more dimensions (such as described by Kline, 1985) will defy comprehension or representation.

Just as a three-dimensional object can have radically different appearances in its various two-dimensional projections, as was seen through the examples of standard views shown in the previous chapter, an object of more than three dimensions must be expected to have multiple and different projections. A single projection of a hypercube was shown in chapter 3 (see figure 3.5); however, many other views are possible. Without knowing the rules of their projection—orthographic or otherwise—into two dimensions, it is impossible to synthesize these limited views into the comprehension of any whole. As yet, we lack a system for assembling the projections of four or more dimensions into three that takes us very far toward understanding higher dimensionality. As long as this is the case, disagreement over what space is and how it ought to be represented must be expected. The likelihood that any two viewers will choose the same two or three dimensions or planes of representation (out of the 10 or more possibilities) is statistically minimal. Even if two or more viewers happen to share the same reference space (or dimensions), and agreement might be said to exist, this coincidence may well overshadow the fact that such an agreement is always rooted in a limited perception of a space of higher dimensionality that is difficult to recognize. We simply lack the cognitive tools to perceive either the consistency or the incompatibility of spatial representations (projections) as the consequence of their construction within different and limited reference systems. Disagreement over practices and representations of space, and over spaces of representation, is a nearly certain result.

As a model for representing and reconstituting partial projections of subject matter that actually exists in a space of higher dimensionality, graphics, especially orthographic drawing, is more useful than either words or numbers. But we don't yet have an appropriate label, such as "literacy" or "numeracy," for fluency in visual communication practices equivalent to fluency in graphic design or drawing. However, lacking a proper name for a discipline of "the visual" is ultimately not much worse than having the wrong one.

The often-used term "visual literacy" is, as Mitchell observes, an unfortunately limited and biased metaphor, which suggests that visual experience might be explicated on the basis of a language model. Pictures cannot be reduced to words or numbers. There is always something more, something left out, that is unaccounted for by language or mathematics. Being satisfied with naming and quantifying things is a way of hiding this "something more." But the remainder never really goes away. If we are to "understand" pictures, it cannot be through words or numbers alone or even in combination. Pictures can't "speak," which is a function or orality; they cannot "describe," which is a function of script; but they always "depict" something. If we were all as concerned with the importance of being able to "read" pictures as well as we read books, we might arrive at a discipline of visual experience on the order of literacy and numeracy.

Such a discipline could be based in a transformation of graphic experience through senses other than vision, just as language was transformed from an auditory medium to a visual one by the invention of script. Perhaps cinema and video, which tie together sound and pictures in ways heretofore unavailable, present us with a starting point analogous to that of pictographs in the development of script. Even though graphics are conspicuously visual, they are not "purely" visual. There is no purely verbal, mathematical, or visual communication; each mode implicates the other in a complex parabolic web of knowledge and experience, and none can be completely explained through projection to another. So far, theoretical attempts by Mitchell (1994) and distinguished art historian and critic Erwin Panofsky (1991), who revived the discipline of iconology around the turn of the century, have called our attention to a set of problems that are not manageable within existing disciplinary perspectives. For example, Mitchell notes that even the term "iconology" used by Panofsky carries with it an implied subordination of the *icon* to the *logos* that interferes with our understanding of what pictures are, what they do, and how they work (24). Even though pictures often inspire, and even demand, discussion and criticism (which is carried out through language), these activities reify the privileging of language in the (parabolic) construction of meaning more than they illuminate the subject of pictures.

Mitchell suggests that "iconography" is perhaps a better term than iconology, but this too carries excess baggage. An "icon" is an object fixed in space; it stands for something other than itself and it usually embodies authority. Its naming establishes a parabolic relation that tends to overshadow its graphical character. Any map or graph of the internal and external relations that make an icon "work" may be explicable through words, but knowing these relations tells us far more about how they are instantiated through and in language than about how they were conceived and made in the first place. "Theorizing" pictures through their translation into language, which is central to the project of art history and criticism, is ultimately self-referential.

Mitchell's suggestion that we "picture theory" is a limited but useful step toward understanding the power of pictures in contemporary culture, but it is far more difficult to execute than it is to conceive. As Davies (1990:32) observed, even though over the last 50 years or so pictures have become perhaps the main means of the ordinary person's "reading," not many of us know how to "write" pictures. Without such knowledge, culturally and theoretically speaking, the theory and practice of graphics remains *terra incognita*. To understand how pictures "comment" on themselves and objects or events in the world, or to understand how words, numbers, and pictures work in concert, we have to study pictures, in addition to words and numbers, on their own terms. To translate them into some other mode of communication is insufficient. Even though words are unavoidable, we must be extremely careful in our choice of language in approaching the problematic relation of words, numbers, and pictures, since the names we use shape our perception of the objects and relations being named, at all levels of thought. As long as we perceive pictures to be an unnecessary or at best optional means of communication,[5] what they contain—that knowledge which resists containment in language—will remain unavailable. Besides acknowledging the ultimate "incommensurability of language and vision" (Mitchell, 1994:64), we need to understand how and why pictures work, both alone and in conjunction with words and numbers, in the production of social space.

The history of space so far has been written in both words and numbers through the geometric art and science of perspective. The historical and cultural significance of perspective goes far beyond its obvious and visible utility as a graphic tool. Perspective, as I argued earlier, is more than a technique for the graphic representation of objects in space; it is also a model for thought that now pervades the mentality of the West. It is deeply and parabolically embedded in the linguistic and logicomathematic practices associated with the promotion of science and technology over the last several centuries. Confirmed by the apparent precision of the monocular lens of the camera in photography, film, and video, graphic perspective now appears to be a fact of nature rather than a human invention that is specific to our own cultural place and time. However, perspective does not really constitute a language as do the symbolic notations of script, print, arithmetic, algebra, or calculus. The symbolic form of graphic perspective has a fundamentally different character than the symbolic form of true language (Panofsky, 1991). Although perspective might constitute a discipline of vision, it is perhaps too complex and difficult a practice in its "pure" graphic form to ever gain widespread practical currency by itself. This disadvantage has little impact on the deeper effects of perspective, which are demonstrated in social practices and beliefs. In its nonpictorial forms (the script of language and mathematics), perspective is nearly invisible; it cannot be seen for what it "is" because the relation of observer and field of view is abstract, and it can never made as visually explicit as it is in pictures. Beyond the two dimensions of paper inscriptions, perspective is inscribed in social space as a cultural standard. Indeed, as a social practice it can be considered a form of *con*scription into the invisible Panopticon of perspectival thought and action. Here the illusion of centrality and individuality both guarantees one's ongoing performance as a Foucauldian "docile body" (Foucault, 1979) and obscures the cultural script that is being played out. Perspective is neither natural nor neutral; as it appears to work *for* us it also works *on* us. It is not only surveillance; it is discipline.

The perspectival observer, from a high enough point of view, surveys the entire world from what is conventionally called a bird's eye view, but which might better be called a God's eye view. By giving wings—not really bird's, but rather, angel's wings—to thought, perspective reinforces the illusion of human power and domination over both the built and natural worlds. As perspective organizes and unifies the visual and pictorial field around a single point in space (the "eye point" or "station point"), it also resurrects what Gombrich (in Snyder, 1980) calls a "perfectly rational standard of truth" for any (not just perspectival) "arrested monocular image." The "eye-witness principle," a basic principle which Gombrich claims has been recognized (but not always observed in practice) since the Greeks, declares that the work of the artist "must not include anything [that] the eye-witness could not have seen from a particular point at a particular moment" (190). As such, it is essentially a "negative" principle.

However, "what can be seen" at any moment, from any place, is determined not only by physical and spatial conditions but also by the fact that our vision is conditioned through the experience of pictures, especially perspectival representations. The relation between what we see, or can see, and what we can represent through pictures or other means, is "a reciprocal one" (Snyder, 1980:234). Our vision is conditioned through the experience of language and the perspective of literacy, but the reciprocity of vision and language tends to be obscured by what Mitchell calls the historically problematic relation between words and pictures.

Maintaining the (artificial) distinction between language—as demonstrated in talking, reading or writing about pictures—and pictures—as demonstrated in either making or viewing them—must be recognized as a key step in maintaining the privileged status of language as the basis of constructing meaning. Ultimately, this distinction must be discarded as counterproductive; it serves only to lead one astray into an infinite reciprocity, a regressive loop. This is not to say that one must discard language itself; it is only to say that the (parabolic) relations between words, numbers, and pictures must be redefined to arrive at a comprehensive view of the meaning that they collectively describe.

Re-visioning Space

The field of design may serve as the basis for breaking this loop, even though design cannot yet be said to constitute a real discipline. It is, rather, diffused across multiple disciplines, perhaps most visibly, in graphics, engineering, and architecture, but also, in less conspicuous form, in all technological activities from chemistry to computer science and even in textual disciplines such as communication and composition. Ironically, even though "composition" is conventionally associated with the discipline of writing, it is etymologically a synonym for design.

The language of most design disciplines is drawing, but one must use this verbal metaphor carefully. As Mitchell points out, pictures may not be amenable to explication through a verbal model. "Visual literacy" is as oxymoronic as "oral literature." Such metaphors, as Mitchell again points out, serve primarily to subordinate the visual (the *icon*) to other modes of thought (the *logos*). The study of drawing alone is not enough to develop its full potential as a consistent language for the exercise of the visual and spatial imagination, that is, thinking geometrically. The visual sense must first be liberated from, and then recombined with, both language and mathematics to form a new model for thought.

The invention of perspective 500 years ago provided artists and scientists with a precise and consistent language of vision. However, knowledge of this language is, by and large, still restricted to professionals and a few amateurs; the rest of us are no more than consumers of their productions. Perspective no longer serves our needs. But for any other visual language to emerge, to subsume perspective and to affect a larger sphere, it must be widespread and accessible to the extent that ordinary people not only consume but produce it. Drawing, especially orthographic drawing, because it is accessible and usable on both a practical and theoretical level, may serve as the ideal basis for developing a language of vision and, ultimately, for producing differential space.

Not all spatial practices leave traces or representations that can be studied at a later date, and not all practices have a formal structure or pattern that constitutes an explicit "space of representation." Scholars of visual communication, from art history to film criticism, tend to focus on exemplary objects such as paintings; I believe that the study of the practices involved in visual production will tell us much more than these objects do about how people actually see and live space. Social space, and the objects and relations that exist in it, can only partially be represented in any work, artistic or otherwise, which is always an act of reduction into fewer dimensions: usually the two dimensions of print or the picture plane. The manifold, nonlinguistic and nonmathematical dimensions of social space resist full expression in either words, numbers, or pictures.[6] These are products, static and fixed in space; to arrive at an understanding of how they are produced and how they can produce changes in the mode of space, we ought instead to examine the dynamic and mobile practices involved in the production of this documentary evidence.

I will therefore focus on several broad areas of spatial practice that have produced solid documentary evidence throughout the history of Western culture—language, measurement and calculation, and drawing—and are recorded as literary, mathematical, and graphical texts (that is, representations), respectively. These constitute the three major dimensions of our contemporary space of representation. This space is produced by the systems that govern these practices: at one level, grammar and syntax, arithmetic, and various forms of projection in drawing; and at another level, encyclopedias, libraries, financial records and museums, as well as the ways in which people use these systems and knowledges.

For several thousand years, language has been the most dominant member of this triad of spatial practices. The history of language demonstrates a progression that Ong (1967, 1982) calls the "modes of communication": oral, literate (script and print), and electronic. These are roughly parallel to the progression of Lefebvre's modes of space. Absolute space, the space of preliterate culture and early literate culture (before literacy became widespread), is distinguished by the linguistic condition that Ong calls "primary" orality, where face-to-face, oral/aural communication shapes a culture untouched by writing. As literacy spreads through the development of script and print, culture enters the mode of abstract space. Primary orality is as inconceivable to the literate individual living in abstract space as literacy is to the preliterate. Oral patterns of speech and thought persist in the literate individual, but are transformed, to constitute "secondary" orality in the mode of abstract space.

The relation of abstract to absolute space, like the relation of literacy to orality, is one of transformation rather than replacement. The relations of humans to one another and to the earth ("consanguinity and soil") are not lost entirely, but rather are redefined or subsumed by other relations in abstract space. One of the crucial requirements for "living" abstract space is being able to read and write, that is, to be literate. The status of "literate," as opposed to "illiterate" or "preliterate," is a demonstration, as Barry Sanders (1994) points out, of how literacy is more than a way of seeing and knowing the world. It is also a means for differentiating, stratifying, and organizing social space, even among those deemed literate. In oral cultures one consults with elders—the holders of knowledge in an oral culture—to answer important questions such as those of heritage or property. In literate culture, one consults the documents held by churches and governments. In a literate culture, knowledge is located outside individuals (on paper and now electronic media). One's position in literate culture is determined by how much power one holds within the documentary information structure. It's no surprise that anyone who has "retired" and is no longer active in the production and organization of information holds little value to those who are active in it. Elders, along with considerably old information are, in a word, obsolete.

This is but one example of how abstract space "relates negatively to that which perceives and underpins it—namely, the historical and religio-political spheres" of absolute space. At the same time that the inscriptions of abstract space accumulate ever more history, these same inscriptions are continuously redefining this history. However, abstract space also "functions *positively* vis-à-vis its own implications: technology, applied sciences, and knowledge bound to power" (Lefebvre, 1991:50). To manage the inscriptions accumulated over time, better "tools" (like computers, for example) need to be devised to sort, organize, and otherwise control what quickly becomes in Latour's words (1990:40) a "cascade" of information.

The opposite (negative) yet cooperative tendencies of abstract space can also be seen at work in a variety of ways. For example, literacy becomes the exercise of cultural power through the simultaneous leveling of cultural differences and the fragmentation or erasure of traditional social structures through the imposition of a standardized language and grammar. Similarly, a belief in progress as the inevitable growth of technology also levels the differences between traditional cultures.

The tensions and contradictions engendered by and harbored in abstract space inevitably and continually accumulate and multiply. These contradictions, which are perpetuated through multiple and unequal instruments such as systems of public education and the subsequent economic privileging of educated elites, eventually become unmanageable, often to the point where the exercise of economic, political, and military power is required, each instance of which only adds to the set of contradictions that must be dealt with.

Differential space promises to "restore unity to what abstract space breaks up—to the functions, elements, and moments of social practice" (Lefebvre, 1991:52). This promise might be construed as a return to, or a restoration of, our cultural roots in absolute social space. Construed thus, electronic communication technologies are understood to play a role analogous to the role of poetry in oral cultures, which Havelock (1963) called the "tribal encyclopedia." The global diffusion of electronic communication technologies contributes to the reconfiguration of space in terms of global domination: this too suggests a return to the undifferentiated character of absolute space. After all, to the extent that we are all involved in technological culture, we are all consanguinary. The entire globe has been measured and mapped by satellite technology. We can travel anywhere in the world within a matter of hours without really getting lost; with electronic location and navigation systems, we can determine position anywhere on the globe within an accuracy of several meters. However, the appropriate cultural and physical conditions—that is, the space(s) of representation—for producing absolute space no longer exist. Especially significant is the fact that we are no longer attached to the land in ways that the ancient Greeks were. Physically and geographically, our technological culture has left no natural or human space on Earth untouched: detritus—like empty pop and beer cans—and apparatus—like transistor radios, broadcast networks, and cellular phones—now litter the world. Consanguinity without soil does not support a return to absolute space.

Exactly how the "unity" of differential space is to be accomplished is anything but clear, particularly given the tendency of both absolute and abstract space to reproduce themselves. Rather than presenting any explicit path or plan, Lefebvre instead calls our attention to the increasing weight and complexity of the imbalances and internal contradictions of absolute and abstract space. He does not offer a detailed description of, or a plan for, working toward differential space, but he does speculate regarding a range of possibilities for change. Because we do not yet have a science of space, the outcomes necessarily take the form of speculation; the future is not entirely predictable. The question Lefebvre demands that the reader ask echoes my own own: Do we really want to live out our lives in the confused and violent flux of "contradictory" space, or can we work our way through to something better? Without a science of space, that is, a "quasi mechanics of history," our prospects are dim in this regard. In their current form, neither words and numbers nor pictures serve adequately as a visible and consistent language that will allow us to break with our confusing past. It's not enough to just deconstruct language, mathematics, or graphics. Something must be reconstructed from their rubble if we are to manage to communicate at all.

The cultural and disciplinary divisions between the modes of representation necessary to the production and reproduction of abstract space present substantial obstacles to our developing a better understanding and means of producing differential cultural space. One obstacle in particular seems nearly insurmountable: our dependence on language. Language is a clumsy tool for addressing problems of space; its use tends to preserve the dualistic and contradictory character of abstract space. As long as we depend on it as the dominant medium of thought and communication, our ability to envision and enact cultural and spatial change will necessarily be limited. We will not so much produce differential space as we will reproduce abstract space, replete with its imbalances, contradictions, and conflicts.

Perhaps the "unity" we seek can be achieved by acting on the the recognition that words, numbers, and pictures all have something significant in common: They are all parabolic forms of visual representation. Their dependence on the faculty of vision is a potentially unifying force. After all, it is not the faculty of vision itself that distinguishes and separates modes of communication (as Ong and others would have it). Rather, it is the subordination of the visual sense to literate thought patterns that distinguishes and separates modes of space. The faculty of vision needs to be liberated from its cultural and intellectual enslavement to literate and numerate modes of thought if we are to further develop the faculty of envisioning, that is, designing, new possibilities that lie outside the boundaries of the literate and numerate imagination.

We must resist the ways in which pictures are bound up with language and numbers in a complex subordinate relationship. We must resist translating pictures into verbal or mathematical form, or prioritizing the verbal or numerical in examining the relation of pictures to words or numbers, recognizing that to do so is to lose sight of their difference. This does not require an outright rejection of literacy and numeracy or the isolation of pictures from language and mathematics. Indeed, the study of textual, mathematical, and graphic objects in isolation from one another cannot lead to an understanding of the dynamic and multidimensional character of the milieu from which they emerge, or what they might lead to. We need to understand spatial practices and their products, including space itself, not so much as they are written, quantified, or depicted—as distinct representations in the mode of abstract space—but as they are lived in the mode of differential space. As Duncan Davies, in *The Telling Image* (1990:153), observes: "We need the best of the verbal and numerical past, but it must be pliably united with the pictorial opportunities of the future." Graphics—pictures—in conjunction with words and numbers, can open up the imaginative construction of a multidimensional space. By envisioning our task as a problem of space, we might contribute to a flowering of the historical sciences and the development of a science of space. But most important, we would be participating in the imaginative construction of multidimensional knowledge and a reconfiguration of the spaces within which we live. At best, this constitutes a powerful tool for designing the spaces in everyday life.

CHAPTER 6

PRODUCING THE MODE OF DIFFERENTIAL SPACE

The assumptions and the goals of history and science may appear substantially different, or even at odds with one another, but what they have in common is the notion of space as something independent and outside human existence. Even though the physical space of the universe will no doubt continue to exist whether or not human beings are part of it, any concept of space, like any other concept, must be understood as a human (and therefore parabolic) construction. As such it is never more than a partial, reduced, and crafted projection of something too large to be grasped in its entirety. Space is inextricably bound up with the human aims that are served by its invention; we cannot see, comprehend, or communicate everything there is or might be known about our own immediate world, much less the entire universe. Furthermore, we realize, thanks to modern physics, that the mere act of observation changes what is observed; measurement and understanding of what we have observed requires that we observe—yet again—what we have changed. And in the process of our investigations, we—the observers—are also changed. From a microcosmic to a macrocosmic level, investigations into both physical and ideal space have periodically led to the revision of our concepts and practices of space.

One can no longer claim with confidence that space is unchanging, absolute, or independent of human existence. Our conceptions of space always affect how we see, understand, and act, and this in turn can either contradict or reproduce our cultural conceptions of space. If space is, as Lefebvre observes, a social product, a true "science" of space may be impossible. What science (perspectivally, at least) tells us is that accurate observation requires a certain amount of distance or isolation between the observer and the observed, and the outcome of interactions that cannot be observed without interference cannot always be predicted with certainty. Since we cannot ultimately distance ourselves from the objects, phenomena, or events—from the extraordinary to the ordinary—that occur in that space, we are doomed to relate to that space with incomplete, partial, conflicting, and unreliable knowledge. Doing "the right thing" in a difficult situation, when one is faced with uncertain knowledge and unpredictable outcomes, involves best guesses and has to be based in conscience. For example, Jason Bourne, the prototypical geometrical thinker, was ultimately rescued from causing his own demise because he was able to allow the emergence of his conscience (embodied in the personality of David Webb) at the last possible moment.

At a social level (in our own democratic culture, at least) the exercise of conscience means that one has the right to exercise his or her own freedom if it does not infringe on the similar rights of others. At an individual level, the exercise of conscience means that one must behave consistently with one's own principles, which usually resonate with cultural values. This is not always easy to do. A problem we all face in the course of our personal and professional lives is that we are routinely confronted with choices that are incompatible with our values. Measuring success in these confrontations is difficult; nobody always gets it completely right. Perhaps the best we can hope for is to live comfortably with our decisions and their consequences. Whether or not the exercise of science can guarantee a desired outcome in a given situation, the simultaneous exercise of conscience at least guarantees that one can live morally with whatever the results may be.

Conscience sometimes seems to be a rare commodity, and its exercise may seem irrelevant in comparison with the exercise of other apparently unconscientious forces (such as politics, economics, and technology) in our lives. As long as we think of space as something outside our own selves, it is impossible to recognize our ongoing participation in the production and reproduction of social space and its contradictions. Our bodies and voices move in space; our works traverse the globe through wires, radio waves, automobiles, ships, and planes. The physical and social forces that act on us and compel us to act and think the way we do have a spatial component that all too often escapes our notice. But the characteristic spatial behaviors, values, and interests of any culture define the dominant "mode" of space. Although it is not typical to conceive of one's values as based in a prevailing mode of space, or of one's actions as spatial practices shaped by that mode of space, the concepts of "modes of space" and of the social production of space have great explanatory power.

We have yet to develop an appropriate cognitive model for the emergent form of differential space in the sense that geometric perspective is the model for abstract space. Because we live primarily in abstract space, we are hampered by its requirement for a literal and figurative distancing of the observer from the observed. Being too close to the object of observation creates intolerable distortions, as can be seen in any photograph taken with an extreme wide-angle lens. We are still too close, and that closeness distorts our ability to envision alternatives.

Continuing to rely on the obsolete model of perspective prevents a practical or theoretical understanding of the character of differential space. The Greeks, despite their remarkable accomplishments, were ultimately limited in their progress by their lack of the conceptual tools (such as those that perspective was to provide) needed to integrate their compartmentalized and unsystematic knowledges. By insisting on the exercise of perspective under conditions in which it really doesn't "work," we are similarly limited. We must move beyond perspective if we are to arrive at an understanding of how words, numbers, and pictures work together to produce multimodal knowledge. "Putting things into perspective" is always a monocular reduction of spatial relations and differences into two dimensions, and what is shown at all depends on the selection of position, distance, and view direction. As the basis for a unified geometrical space, perspective insists on reducing the multidimensionality of space to a singular view and picture plane. A perspective is always a representation of fewer dimensions than those of its subject matter. The third dimension of a scene—depth—is *always* missing from the two-dimensional plane of inscription, but it can be restored through "reading" the relative size of objects and the convergence of parallels to a vanishing point. What cannot be restored is the detail that has been eliminated during construction of a perspective. Through an effort of imagination, we can recognize objects that have been blurred or distorted (such as distant or obliquely viewed objects), but no amount of imagination is sufficient to put something in that is not there to begin with; what we end up doing is putting ourselves into the picture instead.

To be effective geometrical thinkers and actors in the multidimensional framework of differential space, we need to be simultaneously "in the picture" and "outside the picture." If we want to see as much as possible (true omnividence being an impossibility) we need to be in many places and utilize multiple modes of knowing. We need to find the basis of how these multiple and partial spaces project onto one another.

Descriptive geometry, as a model for putting together pictures of multiple dimensions, is an appropriate framework with which to envision knowledge of the multiple dimensions of social space. Orthographic projection, guided by descriptive geometry theory, largely avoids the distortions introduced by perspectival projection, the very distortions that provide the optical realism of perspective drawings. Orthographic projection works equally well in any view direction, unlike perspective, which works well in one direction only. Because of this multidirectionality, descriptive geometry and orthographic drawing provide a basis from which to envision physical spatial problems involving objects, processes, events, etc. Most important, they provide a parabolic basis for identifying and solving problems of abstract space, for making retrospective projections—historical "pictures"—and future projections—design "pictures."

Documenting the constructive and imaginative assembly of multidimensional space draws attention to the multiple alternatives that are always possible. Life often seems random or subject to chance, but it is nonetheless the outcome of various and often competing human activities that work to shape relations among and between the earth, events, people, and things. The processes, organizations and institutions, and physical objects—from simple to complex, from hand tools to virtual machines, from rudimentary shelters to cities—that are dedicated to establishing and maintaining stability in the world are all, for better or worse, *designed* both in and as multidimensional spaces.

Understanding the multidimensional character of "inscriptions" (and producing these inscriptions) and their translations while we are in the midst of emergent (both real and virtual) space(s) requires a knowledge of space as multidimensional, which in turn requires the development of nonperspectival modes of spatial theory and practice. It also requires a restructuring of traditional and emerging patterns of inscription and knowledge based in the established power relations of abstract space. For example, disciplinary distinctions between word and image, art and science, and designer and user only make sense from the "perspective" of abstract space, that is, the planar space of perspectival and textual inscription and parable that projects our experience of the world and ourselves into abstract space.

The emerging field of design studies calls attention to the designed character of this activity. It opens up the study of design on both a theoretical and practical level by conceiving design as "a diverse unity of all the arts of making" (Buchanan, 1995:43). Design, in this architectonic sense, potentially constitutes a new and distinctive liberal art. From the point of view of design studies, it may be possible to reconceive language and mathematics as design arts (or even sciences) and to understand better their parabolic relation to graphics as technologies of cultural representation and transformation. Turner (1996) provides a suitable starting point for this project in identifying the parable as the basic unit of thought. Parables, as they become more complex, are organized in formal structures that render thoughts comprehensible. Language, mathematics, and perspective are formal parabolic structures. So too are descriptive geometry and orthographic drawing, which, unlike perspective, insist on understanding the shared, underlying parabolic character of words, numbers, and pictures as different means of solving spatial problems through the imaginative and constructive processes of design. Investigating the multimodal role of drawing (and drawing theory, especially that of perspective and descriptive geometry) in design is not only a way of answering Mitchell's challenge to "picture theory." It is also a way of redefining the role of both picturing and theorizing in the parabolic construction of knowledge. A prerequisite for such a project is to begin to dismantle the apparent boundaries between words, numbers, and pictures—boundaries that originate in the disciplinary and cultural divisions that both underpin and result from the production of abstract space. Restoring a "unity" of these modes of representation is a prerequisite for producing differential space. This parabolic unity must develop in our cognitive skill, perception, and action.

Unfortunately, our increasing reliance on of electronic communication technologies—both hardware and software—may hinder the process and work instead to reproduce the divisive and contradictory practices of abstract space. Any kind of manual production of an object requires a mind unified with body, hand, and eye, a process that is discouraged by the mechanical and electronic instruments and processes of mass production. The hand tool—whether a pen or pencil, knife or saw, bow drill or hammer—is a direct physical connection between a living body and a material. This relation no longer exists in the use of more complex machinery like power tools, machine tools, or computers, all of which remove the "operator" even further from physical production processes. When all one has to do is push a button, move a lever, or turn a wheel and the machine does the work, the "operator" only controls the machine. In abstract space, "skill" becomes a task of machine control rather than self-control. Furthermore, by obliterating the difference between word, number, and image—all of which are reduced to "information" encoded as binary digits in "memory" and displayed as "pixels" on the screen—the computer offers us an apparent but illusory unity. With the computer, everyone, from rookies to veterans, is ultimately reduced to the uniform status of "user" except for the elite minority who design the equipment.

Reducing difference by projecting a point of view onto a single plane of projection, and even recognizing and critiquing a dominant view in this manner, is not the same as reconciling difference by understanding how alternate views "project" to one another in a space of greater dimensionality. Our cultural challenge is to project ourselves—through thought and action—into the (social) processes of designing (social) space, to claim design as a democratic activity and to initiate the social production of a new space appropriate to the widespread cultural practice of design: in short, to envision, and produce, the space of difference, "differential" space.

Social space is produced through our actions and patterns of thought and, in turn, our thought and behavior are shaped by our conceptions and experience of space, and the ways in which our thought and behavior project onto one another. If we accept that what we understand as "space" is socially produced, design—not so much as the practical art of "making" but as a formal art of defining and solving spatial problems—will explore not only how space, but society as a spatial phenomenon, is produced. The study of design as spatial transformation is perhaps our best path toward answering Lefebvre's call for developing a "science" of space.

However, at the same time that we need a *science* of space, we also need an *art* of space. I do not intend to eliminate the association of design with "artistic" activity or intention, but rather, to integrate the rigor of design conceived as *planning* with the liveliness and spontaneity of design as *invention*. Indeed, the artificial and problematic distinctions between art and science as well as word and picture are dispensed with in the theory and practice of design as formulated by Buchanan. What the designer ultimately works with and gives form to is the full range of ideas and values that underpin individual and cultural decisions regarding what is meaningful, desirable, and possible. Meaning, desire, and possibility, like space itself, are dynamic and designed, rather than absolute, fixed, or given. They are designed as much or more by group effort than by individual effort, and outcomes are not always accurately envisioned. Whether well or poorly conceived and implemented, the processes of negotiation and compromise by which we arrive at designed constructions are always motivated by and underlie multidimensional purposes and directions. As a result, particular motivations for change and intended goals do not always match the actual outcome. Nobody can possibly foresee (or forestall) all the far-reaching consequences of multidimensional group design. For example, even though Einstein recognized that the theory of relativity could lead to the atomic bomb, he was unable to brake the course of events that led to Hiroshima and Nagasaki. Releasing one's work to the world at large, going public, means giving up control over it, as well as the uses to which it is put, and the results aren't always something with which the individual designer or maker is happy. Frustration and disillusionment are a result of failure on both the private and public sides: the designer or creator (all too often immersed in what designers call "studio mentality") fails to engage in dialogue with the world at large, and the public, which is consequently unaware of the background issues motivating design, all too often appropriates design output for ends and purposes unintended by the designer. As long as the designer's studio or laboratory is walled off from the rest of the world, this situation is unlikely to change. Walls and other boundaries not only keep people *out;* they also keep people *in*. When conventional methods of discipline don't appear to work, spatial separation may seem to be the only answer. It is always an unstable and unsatisfactory solution.

My aim here is to point to a break in these artificial and real barriers. Not only must designers become active and knowledgeable participants in the rhetorical practices of social construction, the public at large must become active and knowledgeable about the various practices and processes of design in order to realize the unique cultural potential of design and design studies in the production of differential space. This project, that of producing or "designing" space, requires a twofold strategy. The first part is to explain what design and the space of design is and could be to both specialists (designers as well as scholars from other disciplines) and the general public. The second part is to explore and develop the potential of the primary "language" of design—drawing—in the project of developing a "language" that is appropriate for a "science" of space and a "quasi mechanics of history."

The study of drawing, as visual representation, has been subdivided among so many different disciplinary points of view—for example, literary criticism, history, philosophy, and psychology—that it is difficult to decide where to begin exploring its potential. However, the dependence of established disciplines on language-based and mathematical forms of knowledge renders most of them unsuitable for this investigation. Starting from any established disciplinary perspective (such as history, literature, or mathematics) entails the risk of remaining trapped in a plane of limited dimensionality and the abstract and contradictory space that is served by its perspective (for example, the "historicism" that Soja warns against). It is through understanding design and drawing as multidimensional *practices* of space, rather than *representations* of space, that we can escape the parabolic limitations of perspectivist thought.

The study of design and drawing as spatial practice is also a step toward understanding better the extent to which we all participate in design. Despite the obvious and pivotal influence of characters such as Galileo, any work of the imagination will be ineffective unless it is understood or at least accepted by others. This acceptation must occur in both theory and practice. The democratization or opening up of design requires both knowledge of drawing practice (the "language" of design) as well as a correction of the stereotypical perception of designers as specialists in the invention and the presentation of image and style; the stereotype obscures their roles in other dimensions, such as in judging what is meaningful, desirable, and possible. We are all designers, makers of our own worlds. Yet without a thorough knowledge of design practice, especially drawing, one can hardly begin to recognize, much less communicate, how one can be drawn into, and designed by, those worlds and other worlds not always of one's own making. Without being "literate" in the language of design, one is not really qualified to participate in design rhetoric, and, without being similarly "literate" in other modes of communication, the designer is hardly qualified to participate in the wider rhetorical context of social, political, or economic debates.

The rhetorical task of establishing the role of design in the production of differential space is a dual process of drawing: both drawing *out* in the sense of extrusion, shaping, and extending our communicational abilities and repertoire, and drawing *in*, or toward, in the sense of making connections between practices of space, representations of space, and spaces of representation. Making the study of design and drawing a cultural priority on the order that literacy once was (and in many ways, still is), is a necessary step toward the production of differential space. To design is to transform social space and adapt one's self to these transformed conditions. Without the consciousness (and, one would hope, conscience) that emerges from a multidimensional understanding of one's involvement in the production of space, we can do little but reproduce the character of the existing mode of abstract and contradictory space.

What I have come to call "thinking geometrically" offers a way of bridging the gap between theories and practices of space through a study of action rather than a study of representation. The main reason I have emphasized the study of action, using the examples of Jason Bourne and the work of Galileo, is that actions are observable in ways that imagination is not, and action, like life generally, is dynamic and multimodal. Action is generally dedicated to the transformation of existing relations; representation is dedicated to the preservation of these relations. Texts, numbers, and pictures are useful but partial evidence, frozen in time; they are two-dimensional, static, and limited representations of selected moments. Language, like mathematics, does not help us understand how to make pictures, what pictures are, or how they work with, on, and for the human imagination. Understanding the power of parable as the basis for the exercise of the imagination in words, numbers, and pictures, and a means of joining partial, multiple, and even dissonant representations through projection, opens up other possibilities.

To investigate these possibilities, we need to study how pictures are made, why they are made, and how they work for and on us. Davies' suggestion for a potential collaboration of word, number, and picture as "pliably united" leads us to an essential requirement in this investigation: a flexible framework, or space of representation, in which to operate. Language and numbers alone or in combination, though most appropriate for solving problems of abstract space, are as they stand relatively inflexible, inadequate to the task of either defining, representing, or solving problems of differential space. Design practice, though based primarily in visual (graphic) modes of communication, also continually makes use of both words and numbers for rhetorical purposes in the design of space. The contemporary designer, whether specialist or amateur, must be proficient in multimodal and multidimensional communication at many different levels. The designer must be a geometrical thinker *par excellence*.

Likewise, we must all learn to think geometrically to answer the questions: How is a new space is produced? How can we direct this production? With this kind of knowledge we can participate responsibly and conscientiously in the design and production of differential space. The production of differential space is doomed to failure, however, as long as we remain in a perspectival framework, dependent on the primary tools of contemporary abstract and contradictory space—words and numbers—to identify and solve what are actually problems of differential rather than abstract space. The defining characteristic of differential space may be ongoing transition, the only constant being change, so to speak. Producing or dealing effectively with differential space requires the ability to adjust one's thought and behavior to shifting and unpredictable frames of reference. No one point of view or frame of reference will suffice; one must be able to make and remake parabolic connections among multiple frames of reference, through language, mathematics, and drawing. Each are translations of each other, and of a larger, more complex and multidimensional entirety than can be expressed in any single projection through any plane of meaning.

Making the transition between language (as it is) and whatever will come next is as challenging for us—in both individual and cultural dimensions—as it was for the ancient Greeks to make the transition between orality and literacy, or for the participants in the Renaissance to move from the Classical to the "modern" mentality. The transition between orality and literacy involves a major sensory and conceptual adaptation on the part of individuals and the culture. Moving from the acoustic world of speech to the visual realm of text—a result of internalizing the process of writing (Bolter, 1991; Ong, 1967)—involves a radical restructuring that is still ongoing throughout much of the world.

Increasingly, much of the world is now moving from the purely visual (and silent) textual world of printed words, numbers, and pictures—the abstract space of perspectival thought and imagination—into the acoustic and visual realm of electronic communication, which promises to transform the production, distribution, and consumption of texts as well as the practices involved in their generation. Hypertext, for example, requires entirely new techniques of both writing and reading. There may be no beginning, middle, or end to a hypertext. Such traditional devices as a table of contents or index—two ways of mapping a book—no longer have the same utility or power in hypertext as they do in a traditional text, because "content" no longer has a fixed or linear structure. An electronic text begins to resemble a database that can be searched, rearranged, and selected as the "reader" wishes; readers thus become the "authors" of their own texts. Reading is much less a silent process today: technologies of juxtaposing animation and sound with text transform hypertext into "multimedia." Learning to make and use multimedia effectively requires different practices of space, representations of space, and spaces of representation.

Given the accelerated flow of information in contemporary culture, it may not take nearly so long for us to take advantage of the possibilities of electronic communication techniques and technologies as it took to develop the cultural potential of print technology, a process still underway. However, I remain wary of many of the possibilities that emerge from the potential application of electronic communication technologies to the design and production of differential space. Technology is a medium, and as McLuhan has argued, the medium is the message. The medium shapes us at least as much as and perhaps even more than we shape it (see Bolter's *Turing's Man*, 1984). Electronic communication technologies do extend our intellectual capacities, but they may more significantly limit and constrain our opportunities for creative and critical thought by reinforcing the dominance of both language and mathematics (which are married in the precise and logical "languages" of computer programming) as privileged forms of knowledge. That is, they may, more significantly, work to reproduce the mode of abstract space rather than enable the production of differential space.

Coupling ourselves with machines—an ultimately dehumanizing relation—is hardly a satisfactory answer to the project of better directing our imaginations and behavior toward the production of differential space. At the same time that developments such as e-mail, bulletin boards, and chat rooms appear to democratize communication, these language-based technologies discipline their participants and reinforce the at-once contradictory and homogenizing tendencies of abstract space: the hypothetical "user" is a part of a "community" (virtual, perhaps, but real nonetheless) that is anything but uniform. It is stratified according to one's level of technical prowess (from rank beginners to hackers and gurus). It is also distinguished economically from all those in the world who cannot afford the space, time, or requisite equipment and who, as the uninitiated, are forced to live with the temporal and spatial limitations of "snail mail," ordinary printed text, and face-to-face conversation. Gaining access to the Internet reinforces the perspectival illusion of growing wings; one can survey the entire world from the God's eye of the computer terminal.

Designing differential social space requires moving beyond such perspectival projection. Words, numbers, and pictures must be recognized as interpenetrating or intersecting planes of knowledge, constituting a dynamic assembly, a multidimensional parabolic space. The distinctions among words, numbers, and pictures, like the distinction between art and science, must be seen as artificial and unnecessary and based in the unexamined acceptance of a particularly limited space of representation—the space of perspective. The isolating and dualistic character of perspectivist (abstract) space now pervades our literate thought so thoroughly that it is difficult to envision any escape from the gravity of its sphere. The study of descriptive geometry and orthographic drawing, conceived as necessary communicative arts and sciences in the context of the "architectonic" discipline of design, projects the possibility of envisioning and producing the mode of differential space.

POSTSCRIPT

BEYOND THE FRAME: ALTERNATIVE PERSPECTIVES ON THE IMAGINED ENVIRONMENT

KEVIN G. BARNHURST

Some time ago I had the occasion to drive with a journalist on Interstate 57, the expressway that goes for miles straight north over the prairie between Urbana and Chicago, Illinois. I was pointing out how in the distance the edges of the road converge at the horizon. "No they don't," the reporter said. "If they did, we couldn't get through." His point was that I misspoke. A stickler for precision, he wanted me to say not "They converge" but "They seem to converge" or "They appear to converge." But I meant what I said. In the distance, the lines converged visibly. The possibility seemed absurd to my passenger because it denies the primacy of engineering over mere appearance. But I pressed forward anyway. Out the windows on either side of my old Mazda I pointed out that the linear edges of the highway were parallel. He agreed that they were, and so I asked, "If we accept what we observe, then how do the lines change from being parallel at our sides to converging at the horizon?" He didn't know. "Do you see any angles?" He didn't. Because parallel lines never converge, I said, there must be gradual changes in the lines, a curvature—surely not angles!—between the places along our sides where the lines are clearly parallel and the place at the horizon where they visibly converge.

We live in a complicated, multifaceted world that presents surprises at every turn, but we are cultural creatures who cannot merely observe. In the act of seeing, we automatically redefine what we encounter as something else. In Europe and America we revisualize things to fit cultural expectations defined by linear perspective, a system codified during the Renaissance. John Waisanen questions that system and proposes a different, geometrical mode of thinking, which he suggests can be tested against "real events" available "through direct experience" (9). This essay proposes several examples to illustrate the weaknesses in the current system and the possibilities for Waisanen's alternative. I first review the definitions of perspective and then draw on several works from art history that seem to clarify geometrical thinking, before returning to some general comments on perspective and its alternatives.

Linear perspective is a technology so thoroughly embedded in every aspect of Western civilization that it seems to be the only realistic view of the world. To replace that dominant mode of thought, Waisanen has set no small task for himself, but one worth the effort. The existing system leaves out "such invisible phenomena as changes in social structures, relations, and forces" (93), which are at the heart of the social sciences. An alternative would benefit not only scholars but society at large. The linear system leaves out other points of view that demand attention, especially in the United States as we begin to recognize that our society is multicultural. What if, when we cross the Illinois prairie, we could see not only the curvature of the road toward the distance and its arc over the surface of the earth but also the ways the road itself imposes lines on the landscape, sweeping out of sight the intervening changes in topography and waterways and local politics and natural history as we pass by?

Systems of Seeing

Perspective is a general term for any system of visual organization used to render the three-dimensional world of human experience onto some other medium, from the flat surface of a newspaper or computer screen to a multidimensional art installation or virtual-reality cave. Several different systems of perspective are ingrained in Western thought and spread through global culture, including spatial hierarchy, atmospheric perspective, and chiaroscuro, in addition to linear perspective. All of them combine design elements (the basic bits of representation, including points, lines, shapes, and space) and attributes (the relations between elements, including scale, direction, tonal value, and color). Cultures use perspective to serve two purposes. One is to establish relative importance. All perspective systems work to express power, usually as a hierarchy. Map projections, for example, are systems of perspective that assign greater importance to some countries or locations. Usually those in power make the maps and place themselves at the center of focus (Barnhurst, 1994). The other purpose is to create an illusion. In the case of maps, the oceans and lands are colored naturalistically, and sometimes illustrations of mountains or ships are added to give the projection a natural appearance. The two purposes of perspective work in tandem. A map's projection asserts what matters, and its surface is then decorated in ways that obscure that assertion. In short, all systems of perspective assign power and privilege but disguise these purposes behind another layer of naturalizing representations, so that the resulting imagery seems completely natural and real.

Spatial hierarchy imposes a discipline of scale, tonal value, direction, and sometimes color to assign greater importance to some objects in space. Assyrian art, for example, used scale alone to suggest rank (a technique called *heirarchic* scaling). The ruler is shown larger than any subjects, and drawings may show slaves even smaller still. In some Byzantine and so-called Oriental art, direction (that is, the relative position of things within the frame or field of vision) appears to suggest moral stature. The venerated personage occupies a position higher than underlings and adorers. In some works the direction (higher or lower) is combined with scale (larger or smaller), and then special colors are assigned to important people. In Christian art, to point to a well-known example, the Virgin Mary wears a blue robe (figure 1). These examples might seem far removed from life today, but in fact all the modern media of communication continue to use spatial hierarchy as a system to organize things for the viewing and reading public. In newspapers, journalists use spatial hierarchy to show the priority of stories: important stories go higher on the front page (direction), have darker headlines (tonal value), and usually occupy more column inches (scale). Web pages play the important pictures and graphics larger or darker or in color. The decisions of the editors and designers in new and old media alike reflect a hierarchy of judgment about the content of news and advertising, as well as about the media producers themselves, their audiences, and everyone else.

The system called *atmospheric perspective* uses tonal value, direction, and in some cases scale to create an illusion of depth, usually on a flat surface. In Egyptian art, one object overlaps another to imply a receding background (figure 2). Roman frescos and vase paintings use gradations in tonal value to suggest receding space. In atmospheric perspective, objects resting closer to the viewer are positioned lower in the frame. Although primarily used for illusion, atmospheric perspective also suggests the relative importance of the objects overlapping anything paler and higher in the image. The system is most commonly found today in photographs, where important objects nearby overlap less important objects in the distance. Although people often assume that photography merely replicates reality, it does not. There is a complicated interplay between the scene and two other factors, the actions of a photographer and the effects of camera designers. The photographer chooses a vantage point that places some things closer and lower down and leaves other things out of the frame entirely. Camera designers developed lenses and related mechanisms that allow some things to appear in sharp detail and reduce other things to a fuzzy blur. Most viewers are unaware of the judgments embedded in the equipment and made by photographers and assume that photographs are simple reflections of the natural world. The system of atmospheric perspective applies outside of pictures as well. Internet sites and display advertising in various media use overlapping boxes, shadows, and tint blocks to create the illusion of additional layers. The use of larger and darker type for more important items imposes a natural-seeming order on the content.

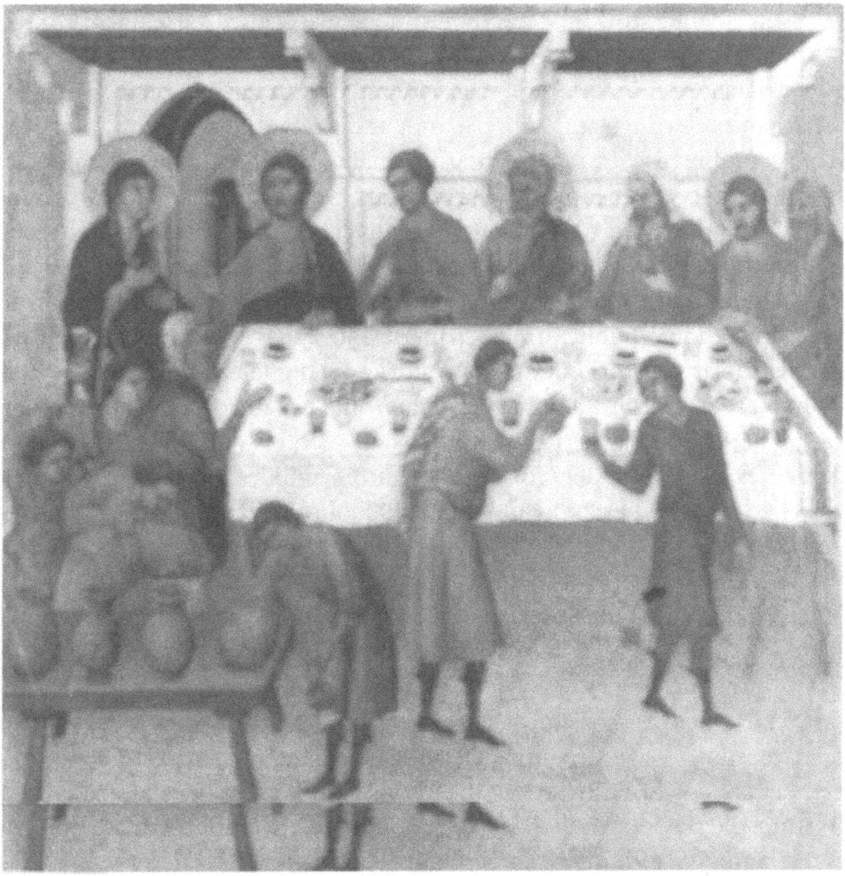

Figure 1. The perspective in this painting by Duccio di Buoningegna (1255–1319) may seem odd to viewers today, but it had a spiritual purpose. The image on the back of a lower panel on an altar piece depicts the important figures as larger in size than their servants and presents Jesus and the members present at *The Wedding at Cana* (c. 1308–11, tempera on wood, 43.5 by 46.5 centimeters; Museo dell'Opera del Duomo, Siena, Italy) through the costume and props familiar to an illiterate public. The purpose of these devices is lost on viewers accustomed to the dominant linear system; instead they create the expectation that items in this painting might slide off the tabletop (see Bunim, 1940).

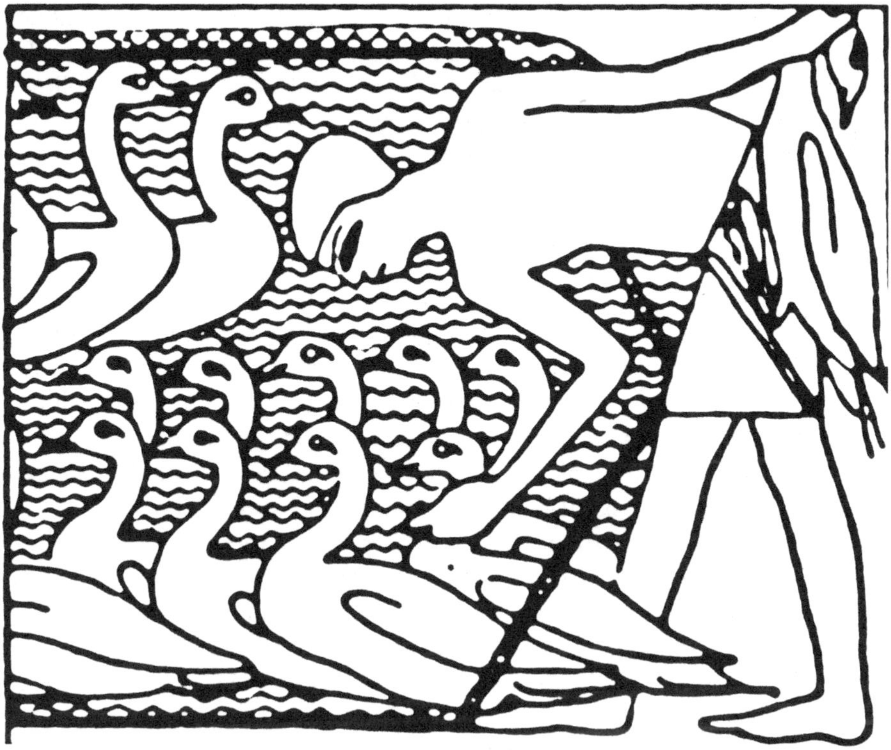

Figure 2. Egyptian renderings used overlapping to place some icons in front of others in contradiction to the flat surface of papyrus rolls or tomb walls. In this detail showing a "Catch of birds being taken out of the net" (redrawn, from the Middle Kingdom; see Schafer, 1974:187), the layering is quite complex; it places the human torso, arms, and hands, shown fully, as if in front, but the hip and right foot are obscured, and the birds are shown in several layers. The effect of depth that results from layering, referred to as *parallel plane representations*, is one of the illusions that combines with others to produce atmospheric perspective.

Chiaroscuro, a third system related to perspective, uses only the value scale to create an illusion of depth. This modeling effect has been used in ancient Mesopotamia and many other cultures since to recreate in two dimensions the play of light and shadow across three-dimensional objects. Greek vase paintings from the fifth century before the current era (B.C.E.) use darker areas to suggest cast shadows, and Roman frescos use both the modeling across the object and a darker shadow cast nearby. Painters in Renaissance Italy developed a system of five identifiable areas or parts, made from the scale of tonal values, to clothe drawn objects in a three-dimensional illusion. The artist begins with a faintly outlined object seen under an imaginary source of light. The *highlight*, at the white extreme along the range of tonal values, is the area nearest the light source. The *deep shadow*, at the other extreme, is farthest from the imagined source of light. In between these extremes is *middle shadow*. To create the illusion of depth, the artist also includes a reflection and a cast shadow as follows: On the darker side, some of the light is supposed to bounce back toward the edge of the object, creating a slightly lighter area of *reflected highlight*. A *cast shadow* is also drawn as a fuzzy area beneath the object and spreading on the ground near it and away from the light source. On a blocky or squared-off object such as a cube, most of these parts of shadow may be separate, but on a curved object such as a sphere or cylinder, they blend in a continuous range from dark to light. Today people encounter chiaroscuro throughout the media. Illustrations for advertising and feature content often employ shading traditionally, but sometimes charts and cartoons violate the rules to produce a comic or ironic effect. Commercial artists use conventionally correct shading to suggest sophistication and naturalism and "incorrect" shading to suggest the primitive, naive, or exotic—anything outside the centers of power.

Waisanen focuses attention on *linear perspective*, the most elaborate of these systems, because it uses geometric relations between lines to produce illusionary depth on a flat surface. It is often assumed that linear perspective emerged abruptly during the Renaissance and that the history of the linear system is all about technique. Neither assumption is correct (Edgerton, 1975). Although Roman art did not use perspective systematically, the architect and author Vetruvius understood the concept in the first century B.C.E. To render a simple building or cube in perspective, the lines cast from its eight corners to converge on the eye are plotted where they intersect the intervening plane of the drawing paper. This rather daunting abstraction was reinvented by Italian painters during the Renaissance and codified by the humanist Leon Battista Alberti in the fifteenth century.

Alberti explains linear perspective not according to its mathematical construction but with an elegant metaphor. His 1435 treatise, *Della pitura*, equates a picture to a view through a window. The metaphor gave the complex arithmetic of linear perspective a commonplace, easy-to-remember image. It had a profound effect on painting by disguising the elaborate machinery of linear perspective as a deceptively simple and seemingly natural way of seeing. The metaphor implies, of course, a particular kind of viewer, who contemplates the world from a particular subject position. The privileged few (usually rich men) in Renaissance culture could view the world in this way, from inside a building, through an ample rectangular window raised above the ground. The metaphor eventually permeated not only painting but also military engineering, theatrical scenery, city planning, and other practical affairs. The demand for mechanical aids to create linear perspective led to the development of optical lenses and eventually to the photographic camera. Then, as photography became ubiquitous, the picture-window metaphor permeated almost every corner of Western society, eclipsing all other modes of vision. The metaphor influenced everything from home and garden design to the invention of tourism and designation of scenic stopping points along highways.

Linear perspective is a system of reduction. It selects some aspects of human vision for inclusion and arrays them in exact geometrical relationships based on straight lines. It renders objects in relation not to many but to a few (one, two, or three) vanishing points. Usually the pattern of lines is obscure, but sometimes computer rendering, used in futuristic movies and in advertising for technical products, does make visible the contours of objects and the base grid of the system. A horizon line establishes the most distant edge of the grid, and then horizontal lines march down from the horizon at increasing intervals to show the approaching surface. Objects resting on that surface have edges that, if extended, would converge on the vanishing points—which one or ones depending on how the object is oriented on the gridded surface. By this means, the volumes that people experience in the field of vision are broken into many chunks (geometric solids based on rectangles and/or trapezoids), and the chunks line up in a limited number of ways. Any rounded objects in this system take their shapes from the chunk they would fit inside or be carved from, and their round edges often become elliptical, so that they take on an egg shape. If this description sounds surreal, then it has succeeded in conveying some sense of what a reductionist and slightly bizarre system linear perspective is.

The linear system manages to hide its awkward construction because it has adapted and absorbed the other systems (figure 3). Chiaroscuro and atmospheric perspective usually accompany linear perspective, and spatial hierarchy plays a role because the important and admirable objects usually appear to be nearer, that is, larger, darker, and lower in the frame. The illusionism of chiaroscuro as well as atmospheric perspective disguises and strengthens linear perspective by mimicking and making reference to the natural world. In other words, linear perspective records the system of power, and other perspective systems reiterate the power structure while making the linear system seem natural.

Although commonly accepted as the most realistic representation of how things look, linear perspective in fact depends on a number of arbitrary choices. Some of these are mechanical, such as the height of the viewing eye and of the horizon, the location and size of the picture plane, and the distance to the vanishing points. When the horizon is high in the frame, objects appear small and beneath the viewer, but objects seem to loom from above the viewer when the horizon is low. Exactly where a picture plane slices through the space between the viewer and the object produces different meanings. A

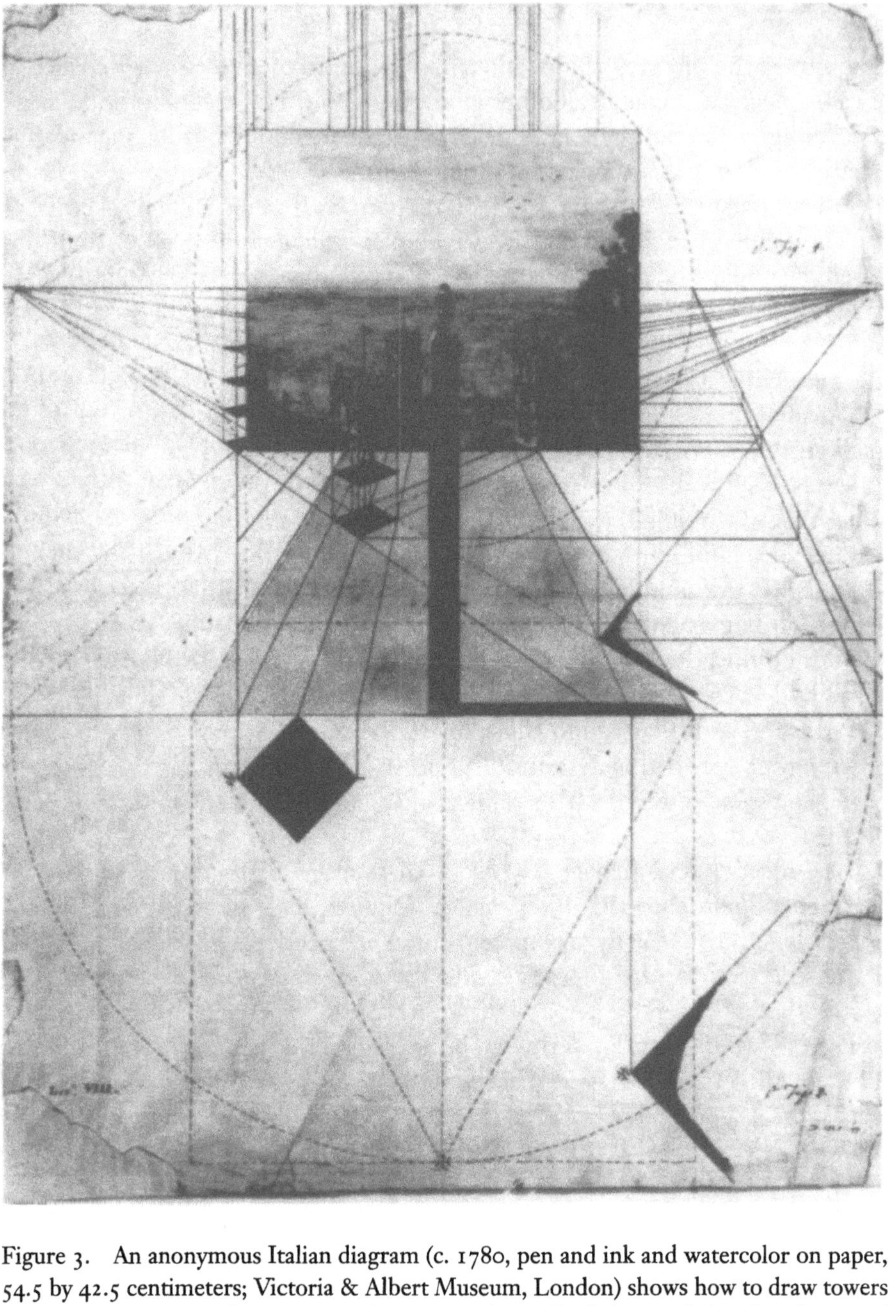

Figure 3. An anonymous Italian diagram (c. 1780, pen and ink and watercolor on paper, 54.5 by 42.5 centimeters; Victoria & Albert Museum, London) shows how to draw towers in a cityscape (see Lambert, 1984). The square, shown flat below and then foreshortened onto the grid in the intermediate gray area, is included within the picture rectangle as a section of one tower. This lesson in linear perspective achieves the appearance of depth by applying chiaroscuro (shading the sides of the towers) and atmospheric effects (overlapping the towers on a landscape of tonal values graduated toward the horizon).

picture plane that is positioned close to the viewing eye emphasizes the surrounding environment, but one positioned close to the object emphasizes its details or blemishes and removes the context. Different vanishing points also produce distinct meanings. Placing vanishing points closer together tends to exaggerate the angles of the perspective lines, producing a distorted quality in the objects that may seem to imply a negative commentary on them.

The public encounters linear perspective routinely but usually without realizing how the images have been manipulated to express power relationships (figure 4). Photographers choose to shoot from a low or high vantage point, and that makes people look small and insignificant or jowly and menacing. Illustrators can move not only the horizon but also the vanishing points at will to express judgments about the world. When designers crop a picture, they shrink the setting, making the environment and context less important, and they enlarge the superficial detail, making people more dramatic as objects of scrutiny. Linear perspective is so universally accepted that production personnel often "correct" a photograph by tilting it until the horizon becomes exactly horizontal, or the posts and corners of buildings become precisely vertical, and then removing the skewed edges (cropping) to align them to the page or screen edges. When few alternatives exist anywhere in popular culture, the standard rectangular picture in linear perspective comes to seem all the more natural and inevitable.

The most important choice linear perspective imposes is not mechanical but philosophical (Panofsky, 1991). Ordinary people do not experience a world with a consistent, horizontal horizon, but instead see the world from a variety of viewing positions. Sometimes the horizon crosses diagonally through their field of vision. They do not usually encounter cityscapes as a grand linear scheme laid out for their contemplation but instead confront neighborhoods through a complex process of walking around them and seeing a few buildings at a time from different angles (figure 5). Their focus of attention may be on one venerated place or one beloved person, upon whom all rays and angles seem to lead. Or they might not even consider the world a collection of discrete objects with lines, edges, or surfaces that signify. Ordinary people from many cultures understand their world as primarily an expression of other things, such as group cohesion, holistic nature, or spiritual reality. In contrast, Renaissance perspective is called *humanist* or *focused* because it places the human subject at the center of the arrayed world, exalting the individual and turning all else into a human domain. Things seen through a window are objects to possess, as a landlord owns his property, and other people are also objects with surface appearances that matter.

The media of communication reiterate this philosophical position daily. The placement of the self at the center has become so widely accepted that any

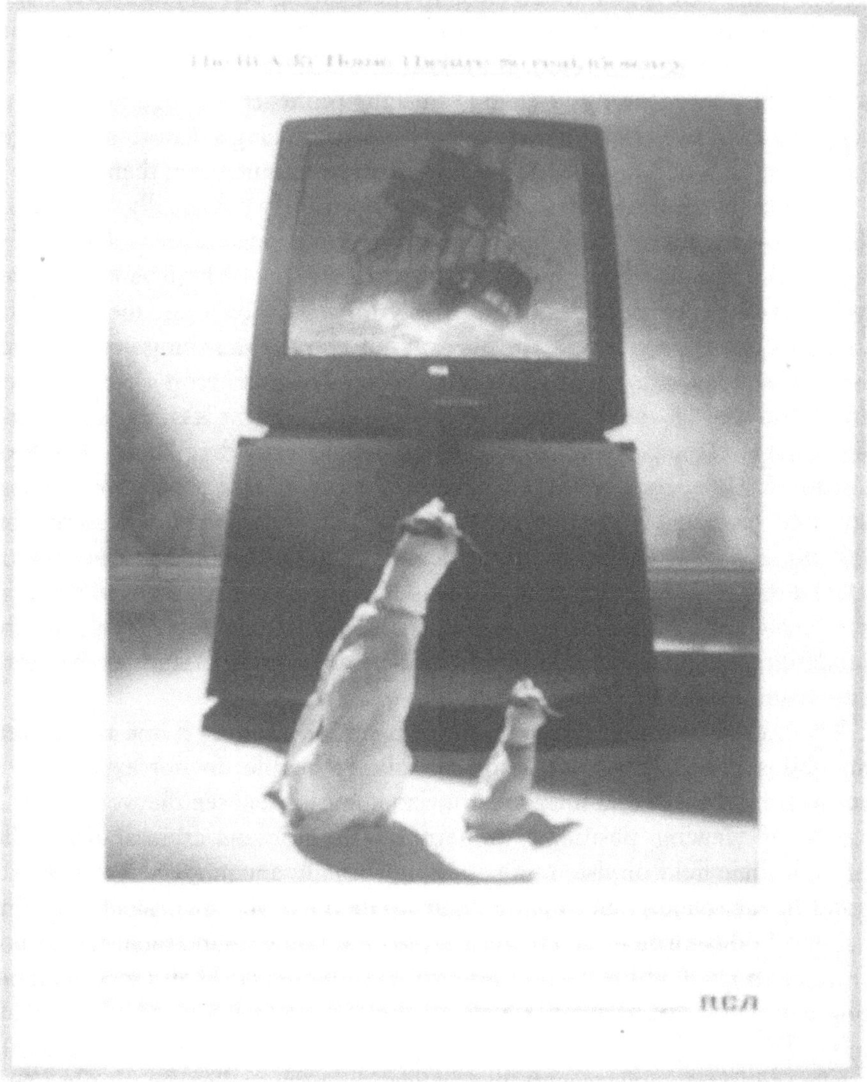

Figure 4. The low angle of the camera and the low horizon depicted in this advertisement combine to make the RCA television set tower over the two viewers. The visual joke of substituting dogs for people plays on several underlying myths. The notion that audiences are easily swayed is applied to others, never to oneself, and so RCA can substitute dogs without offending its target consumers. That sleight-of-hand diverts attention from the controlling myth that technology is all powerful and rightly so.

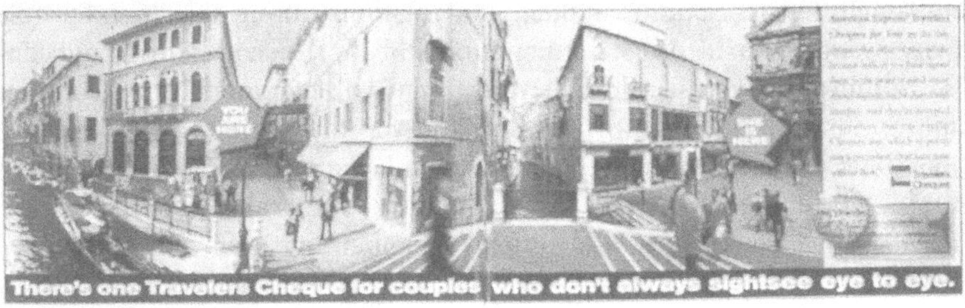

Figure 5. To sell a product enhancement that allows two consumers to sign the same set of travelers checks, an advertisement for American Express employs an alternative system of perspective. The curvilinear system accommodates a heterosexual couple in which a woman is allowed independence, although not without clear reference to stereotypical roles for males and females. By using a curved or fish-eye perspective, American Express can acknowledge its adventuresome customers without offending traditionalists. The company thus seems to celebrate its reach to a faraway place while commenting on how odd the image is of the woman's liberation.

deviation from the ideal can seem grotesque. Artists during the mid-twentieth century consciously broke the rules to produce surrealism, and media images occasionally use the same strategy to produce an eerie or dreamlike effect. The metaphor that Alberti proposed, of vision as what the privileged individual sees through an elevated window, is what Waisanen has examined and found wanting. Geometrical thinking critiques the objective and individualistic assumptions of linear thinking and suggests the need to liberate vision. Waisanen alludes to alternatives that do not define power so narrowly as to make other viewpoints seem erroneous or bizarre and that instead allow for different philosophical choices.

Images as History

Waisanen's hypothesis is itself metaphorical. It imagines an observer simultaneously aware of multiple points of view while witnessing not just objects but subjective and invisible realities. Consider a large entity such as society, for instance. A concrete picture for an entire society will remain elusive, but the metaphor of thinking geometrically might change the narratives people share *about* society. Visual images alone are mute, they tell no stories. Instead viewers create narratives from the surrounding context and from a variety of personal experiences. That process has a hidden dimension, an underlying geometry that structures (and limits) imagination. On the picture's surface, icons provoke memories of a few associated stereotypes, formal properties signal similar cultural associations, and temporal relations among objects and people invoke events from history or contemporary life. These icons, forms, and content within an image also allow viewers to replay the grand narratives of their culture. These four levels of narrative mirror those found in writing: the (sometimes stereotyped) images of objects and characters, the form or plot of stories, the scenes of events they retell, and the cultural values they express. All such narrative devices are also limited by something else—the perspective cues hidden within the imagery—and that is what Waisanen seeks to change.

He suggests that a shift to geometrical thinking would also influence history. Because it combines imagery with narratives about past events important to society, historical art is a good place to begin understanding the problems of linear thinking. When people look at history paintings, they confront a contradiction between their belief in history (what they think really happened in the past) and their unbelief in art (what they consider unrealistic in pictures). Historical art is usually treated as important, but viewers today do not find history paintings particularly believable. Prehistoric art may be almost unintelligible, and ancient art may be too foreign to seem real. Even familiar paintings from nineteenth-century movements, such as realism or impressionism, may seem quaint and out of date. In each era artists have created different styles, different perspectives, and different ways of seeing that not only reflect but also change people's understanding of the past.

Past movements in art routinely reappear today in the style and content of images in advertising, news, and fictional entertainment. As a result, the perspective artists adopt has a broad impact on popular culture and on ordinary people when they imagine themselves in relation to history. As an (admittedly oversimplified) illustration of the ways linear thinking permeates everyday understanding, in this section I describe how my students responded to paintings from three phases in the history of art: the art of prehistoric cultures, the figurative art of the ancient world, and the realistic art that first emerged in the nineteenth century.

While teaching abroad, I had the opportunity to go on a field trip to the caves in northern Spain where artists several millennia ago inscribed on walls and ceilings the images of animals in black and reddish brown (figure 6). Students on the field trip found the images more interesting as anthropology than as art. The drawings seemed simple and obvious, like something a child could do. As we went by bus from the entrance of one cave to another in the hills of the Cantabrian Region, that initial response to the paintings led us into speculative narratives about art maturing from the glacial past. A childhood analogy seemed to present itself: that art developed, from simple to complex, much as children mature. Just as children's drawings depict familiar things, the explicit content of the caves—horses, bison, mammoths, and the occasional hunter—

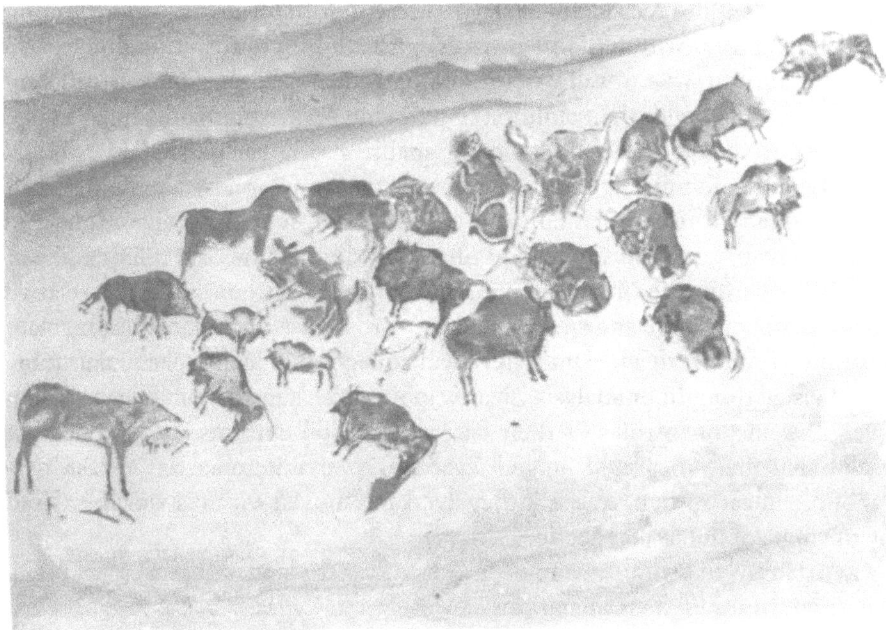

Figure 6. *Cueva de las pinturas*, Altamira, Spain (1998, photograph from the author's collection; see White, 1987).

seemed to reflect the everyday life of a society that sustained itself by hunting. Because children also draw on walls, the location of the Paleolithic works also contributed to the analogy, suggesting that the caves (as stereotyped cartoon images assert) might have been people's homes. The animal icons also tended to call up emotional associations, most of them from stories learned in childhood about things like riding horses, hunting buffalo, and battling for survival—cave dwellers versus the woolly mammoth. These, of course, are related to larger cultural values shared by many young Americans, such as the Darwinian notion of survival of the fittest.

For the students in our group, the form of cave drawings seemed to reinforce the assumptions found in the content. The beasts are arrayed not upright or framed within pictures but in complex relationships with many axes. Such a world may seem topsy-turvy because it lacks a unitary perspective; the drawings are not, in a word, a *picture*. My semester-abroad students expected skilled artists to orient animals vertically and horizontally, placing individuals and groups within a frame (although not necessarily within a literal one). Even without a physical frame, the students expected artists to work within a formal frame that delimits an image area and orders the visual forms and icons within it (they would never have put it that way). From our conversations it was clear that they thought the cave paintings lacked sophistication. As children they had learned—quite reasonably it seems—to color on pieces of paper (framing their crayon lines within defined spaces) instead of on walls, to define clearly what was up and what was down (orienting to the horizontal and vertical), and to draw parents larger than children (employing scale systematically).

These are some of the fundamental qualities of linear perspective already described: the definition first of a picture plane, which is a rectangle (usually), within which all other elements are oriented and positioned according to a central scheme, and the application of scale, which defines the relative size of all objects within the picture plane according to a system that gives relative meaning to both large and small. In other words, my students, who came from a variety of American and some European colleges but had no particular training in visual thought or analysis, clearly applied the standard of linear perspective. The unspoken rules of their various national cultures all dictated that each visual phenomenon be judged according to its adherence to the basic tenets of the linear system. It is as if they lived and moved within a world ordered by the logic of linear perspective.

Later I traveled with a group of anthropology students to the west of Spain, where we explored the Roman ruins at Mérida and Cáceres, in the region of Extremadura near the border with Portugal. Upon returning, I showed the students in my course on critical thinking another image, a reproduction of a painting uncovered in a house in Pompeii (figure 7). Their first reaction was to marvel at that it survived at all. If the city had not been filled suddenly in volcanic ash, the painting almost certainly would not have been preserved. Most surviving Roman art is made of more durable stuff, like the architecture, sculpture, and coins the traveling group had seen in the Roman Museum in Mérida.

The students found confronting an unfamiliar painting from Rome something like sitting through a meeting in a foreign language, an experience that was fresh in their minds. To avoid complete isolation when they moved to Spain, one strategy they tried was to decode what was happening by analogy, not through words but through the context of familiar objects, gestures, and movements. Just as they used those analogies to enter the (to them foreign) culture of Spain, they tried to imagine a Pompeii not unlike their own world. What they could come up with was very limited. The building looked like pictures they had seen of the Coliseum in Rome, and the figures running, gesticulating, and fighting with weapons reminded them of the stereotypical Roman spectacles and gladiators they had seen on old movie reruns.

They were curious to know what story the picture was telling. Art historians, working from contemporary accounts and legal records, have determined that the painting depicts a riot in the amphitheater (Bandinelli, 1970). In the year 59 of the current era (C.E.) a number of visitors were killed or wounded when fighting broke out between spectators from Pompeii and those from the nearby

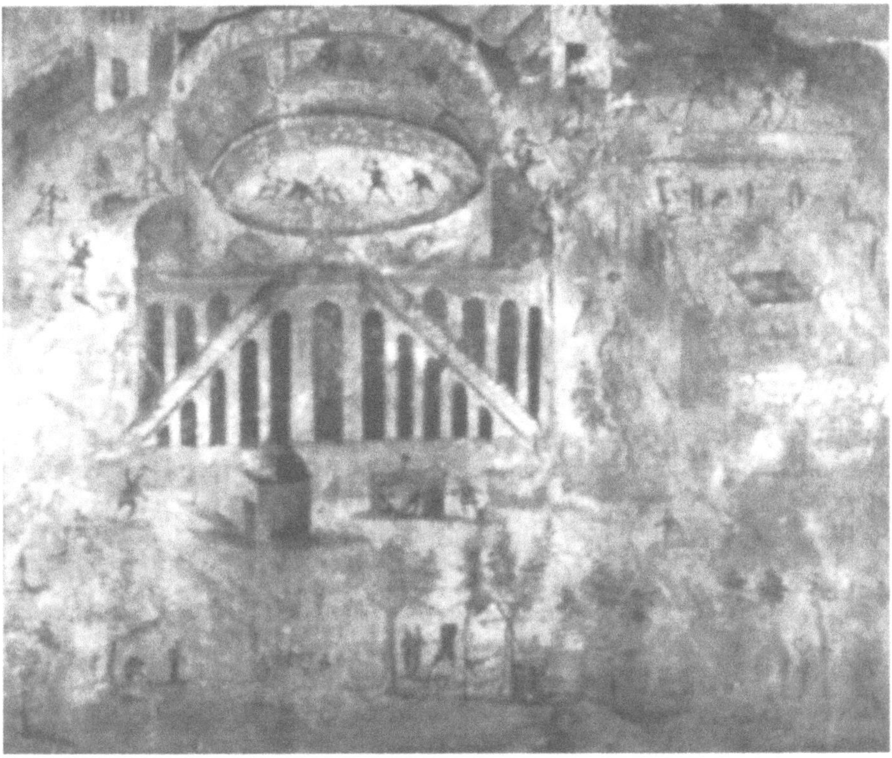

Figure 7. *The Brawl in the Amphitheater*, a wall painting uncovered in Pompeii (c. 59 C.E.; Museo Archeologico Nazionale, Naples, Italy; see Bandinelli, 1970:64).

city of Nocera. As a result of the riot, the Senate in Rome eventually ordered the amphitheater in Pompeii closed for 10 years. That sentence was a harbinger of the eruption on August 24, in the year 79 C.E., when Mt. Vesuvius buried Pompeii in 12 feet of ash, preserving the painted account of the riot.

The students moved eagerly into discussing the picture as a vehicle of narrative. They said they could see how the painting illustrated the events, and they could compare it to familiar narratives today. They recalled soccer matches in Europe that ended in violence, and they said closing a stadium for a decade would be inconceivable today, more drastic than the so-called death sentence in U.S. collegiate athletics. From retelling the events and connecting them to parallels today, it was a short step to the themes my students could draw out: stories of rivalry between two schools, lessons about sport, and narratives about authority and control.

The analogies they drew between the painting and their own lives reveal cultural values. In their experience, individual competition is the primary force behind events. They understand the collective action of teams or fans or rioters not so much as the product of what sociologists call group solidarity but primarily as the sum of individual decisions and personal expressions such as anger. These values are also present in linear perspective, which projects any image of society from the point of view of the individual. In their experience, facts about events are also conveyed through media to individuals. In other words, a picture is not just a projection from a unitary point of view, but any images of public realities are projected back to the individual through media. Linear perspective goes both ways, defining what occurs in the eye (*how* one may see) and defining how the world itself is arrayed for seeing (*what* one may see).

The students in my course did not consider the form of the painting very interesting as art. They criticized the inconsistent way the amphitheater was drawn. The façade is shown from the side, but the interior of the arena doesn't fit its exterior like hand and glove; instead, the interior is seen from a higher angle (but not quite a top view or *plan*, as it is called), and an awning floats detached and suspended improbably above that. My students considered the skill of this Roman artist not much better than the work of an older child today. Some of them said they could have drawn it as well themselves, and they weren't artists. They found the various figures inside and around the amphitheater amateurish. The people appear as simple line drawings or silhouettes, all roughly the same size regardless of whether they are positioned far away from the amphitheater or nearby.

The properties my students found missing from the painting are again fundamental to linear perspective: the rounded objects flattened or foreshortened when seen at an angle, the display of all objects as if viewed by a single eye, the scaling of similar objects to make them smaller if positioned farther from that viewer, along with the atmospheric treatment of more-distant figures as grayer and less detailed. The students dismissed the painting for being at once inconsistent (the arena) and too consistent (the silhouette figures). In other words, the painting lacked a predictable scheme for varying the shapes, positions, and sizes of things. Once again, the art fell short of their expectations, and what they really wanted, although remaining unnamed in our conversations, was linear perspective.

When I asked them to compare the Roman painting to the Paleolithic drawings, they said the cave art was more childlike, and they could identify many qualities that made the Roman painting more advanced. They considered the latter more orderly—the icons are grouped together and the organization of space further clusters them together. The painting, they said, is more clearly a *picture*. In visual terms, the spatial frame orients the icons in the picture (buildings and people) on vertical and horizontal axes. The Roman painting also tells a much clearer story.

The formal qualities along with the recording of events suggested to them that Roman art is more realistic than Paleolithic art. The difference between the two seemed like a clear example of progress, another step in the long trail leading from prehistory through antiquity and arriving finally at today. This commonsense interpretation is also evidence of the cultural values linear perspective can make natural-seeming. If all the myriad objects and events of public and private life come into view in relation to a unitary observer, growing in an orderly progression from the horizon toward the viewer, it would be absurd to assume that history (which is the ultimate presentation of that same array of life) follows some other order, or worse, no picturable order whatever. No, the acceptance of linear perspective requires and underpins the students' faith in progress and organizes their view of the pictures from the prehistoric and ancient worlds.

Some weeks later in Spain, I accompanied an art history professor's class to the Prado, a national palace of the fine arts in downtown Madrid. Unlike the art of less familiar cultures, a painting from the nineteenth century, such as

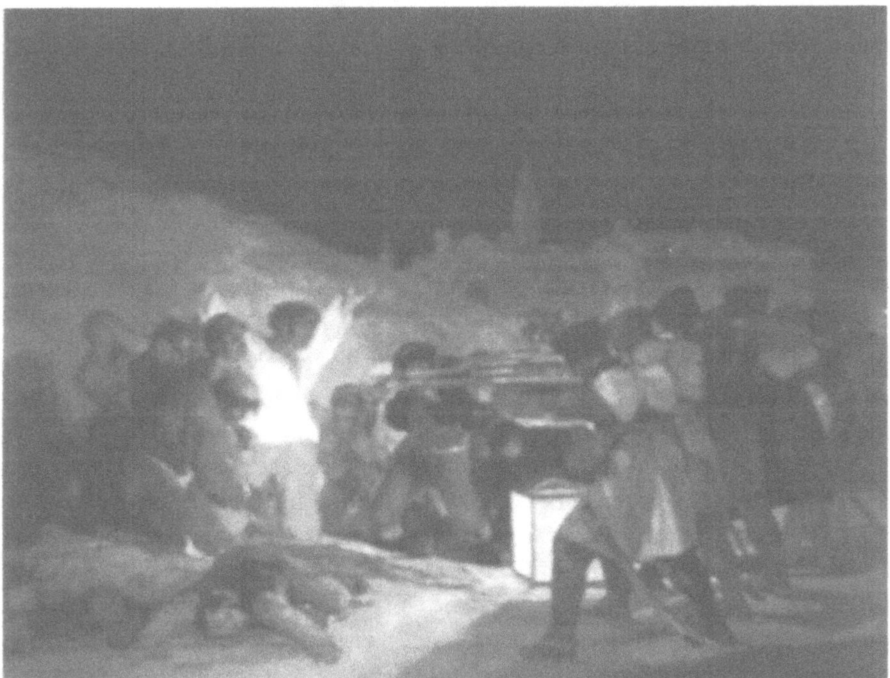

Figure 8. Francisco Goya's *The Third of May, 1808* is also known as the *Execution of the Madrileños on May 3, 1808* (completed in 1814, oil on canvas, 2.66 by 3.45 meters; the Prado, Madrid, Spain).

Francisco Goya's *The Third of May, 1808* (figure 8), appeared to the students less like a mysterious artifact than like a snapshot from history. When I showed a slide of the painting to students in my critical thinking course, the icons seemed obvious: a figure with dark hair and complexion flanked by a friar. The painting brings to mind the stock of Spanish stereotypes, which the students could draw from half-remembered images of the Inquisition, the conquest of the New World, or the Franco dictatorship. American popular culture ties Hispanic-looking characters to the villains of military coups and the cocaine trade, as well as to the well-worn clichés of a passionate Latin temperament. Such images probably colored the students' responses, although their understanding also included positive associations from experiences in Spain.

Even without historical detail, it was easy to draw a parallel to current events, when nighttime executions by firing squads had been in the news from the Balkans. What the students wanted to know was, Who are those people and what are they doing? I explained that French forces had occupied Madrid under General Joahim Murat, one of Napoleon's marshals. The executions took place on May 3, 1808, near the Montaña del Príncipe Pío on the outskirts of the city, after a spontaneous uprising in Madrid the previous day. The distant skyline can be made out on the picture's horizon. Knowing these facts seemed to satisfy the students. They seemed confident that they could understand the painting, that it became fully intelligible once they could identify the events.

Paintings like Goya's seem readily accessible because the audience and the artist share a view of the world. From that common heritage, the painting is clearly a picture meant to convey content: What's happening here? As visitors to Spain, the students also had firsthand experience with the environs of Madrid, and they read the content of the image with relative ease. In other words, because they share with artists of the nineteenth century a common perspective based on Renaissance space, framing, and orientation, it is possible for young adults steeped in Western culture to move directly into the content of the painting (its narrative).

My students praised the painting's form and the artist's skill, but they did notice that Goya painted some passages to look vague or unfocused. By contrast, the more sharply rendered passages dramatize the action of the central figure. They commented that the technique looked more stylized than a photograph. The painting in some ways anticipates the depth-of-focus control of the camera. Photography can render images with great precision and enhance a central narrative moment by leaving other details blurry. Because this is a painting, the students wondered (in a way they would not usually question a photograph) whether it was accurate. Goya did not witness the events he depicted, but he did arrange to visit the site and gather information, much as a reporter might do today. His rendering of events is a reconstruction, not an eyewitness report. The students' discussion of the painting techniques implied that photography would have been the superior witness to the event. This attitude is associated with linear perspective, which proposes the superiority of instruments (the compass and dividers) over the work of the unaided hand.

When I asked the students to compare Goya's to the earlier paintings, they pointed out at once how much more realistic it was than the Roman painting from Pompeii. *The Third of May* seemed by contrast almost a picture of the event itself, made clear by the way the image is organized, with clusters of figures held within a frame that is not only spatial but also wooden. Unlike the cave paintings of Altamira, nothing in Goya's canvas contradicts the order my students considered natural: the horizon line, the regularity of scale, and the orientation to vertical (of people standing upright) and horizontal (falling to the ground). All of my students' observations suggested how much further human progress had come by the time of Goya, and their comparisons to photographs signaled the continuation of that line of progress, toward the direct witness of events that only the high-speed film and portable cameras invented in the twentieth century could achieve.

Multicultural Geometry

I have exercised poetic license in these examples from Spain, both simplifying the events and elaborating on my observations to make a point. The students and their professors did not do detailed analyses of the images as communication, but they did go beyond the immediate reactions I've described to deepen their understanding of anthropology and art history. Initial reactions, however, are the moments when people's unexamined assumptions often reveal themselves most clearly. My point in describing them is to illustrate two aspects of Waisanen's argument: the role of realism in our judgments of the visible world and the role of linear perspective in structuring that sense of realism. I want now to return to the images to begin a discussion of the insights promised by geometrical thinking.

In the case of the cave paintings at Altamira, thinking geometrically might itself shift the analysis to another angle of projection, at least metaphorically, and reveal how complex such prehistoric art in fact is. The notion that Paleolithic art is simple and childlike runs directly counter to another point of view developed among scholars (e.g., Sieveking, 1979), but professional opinion was extremely slow in shifting from the linear thinking mode. The first Paleolithic art to be discovered and recognized during the mid-nineteenth century was *mobiliary* (small enough to be carried). Art experts and historians of the time were startled by the sophisticated decoration of these objects, which contradicted the expectation that primitive cultures would produce childlike art. The fact that the mobiliary pieces were discovered undisturbed among other artifacts in archeological excavations confirmed their age and authenticity. Other works, called *parietal* (fixed on cave walls), were encountered later in the nineteenth century and were rejected for many years as inauthentic. Scholars at the time considered any prehistoric work on exposed walls unlikely to have survived, and of course the sophistication of the cave art contradicted scholars' linear assumptions about progress. Such sophisticated paintings seemed to be fakes produced by more recent artists, who might have imitated the prehistoric mobiliary icons discovered a few decades before. Only later did scholarly opinion come to accept parietal art as authentic work of the Paleolithic era. Scholars could then look at it from a different point of view, one that anyone attempting to think geometrically can now adopt.

A general knowledge of art and design theory can supply another vantage point. The essential codes for rendering a likeness of three-dimensional objects on a two-dimensional surface are fully present in the cave paintings, including the illusions of depth and motion. The quality of line changes within some figures to suggest the weight or mass of the animal depicted, and within other figures the animals appear in the act of running. That the icons are immediately recognizable indicates not so much how easy they are to create (by today's standards) but instead how great an accomplishment they represent (by Paleolithic standards). Contrary to the assumption of progress, the foundation of figurative representation has changed very little in the thousands of years since the cave walls were painted.

Firsthand experience puts the cave art in a different light as well. The works display sophisticated techniques of engraving, paints and pigments, and relief sculpture, not unlike methods used for centuries. These processes required enough specialized knowledge for anthropologists to propose that Paleolithic society had its artists. Certainly the knowledge of representation and technique evident on the cave walls would be beyond most educated nonartists today. I couldn't make anything so durable—to say nothing of the expressiveness and scale of the art—and neither could my students. For me, the experience of viewing the Altamira caves inspired a deep reverence unlike, say, my enthusiasm for the simple drawings my sons used to bring home from school. I had for many years lectured about the cave paintings from slides, but nothing prepared me for the sculptural qualities of the art, which was created to enhance and exploit the curvatures, cracks, and indentations in the natural surfaces. Even the replica at the natural history museum in Madrid could not do them justice. Climbing into the caves was physically demanding, not only because there have been many changes in the surface terrain but also because the paintings are often located deep within hard-to-access chambers. Unlike the cozy image from Flintstones cartoons, these caves would never have made inviting homes.

Field research provides yet another angle to assess the cave art. Archeologists have established that most Paleolithic settlements were in the open air, not in caves (Leroi-Gourhan, 1982). The cave paintings of prehistoric Europe rarely record the berries and only infrequently the reindeer that were apparently the staples of the Paleolithic diet. The technologies and skills in evidence at Altamira and other sites in northern Spain absorbed significant cultural resources, and that investment required a deep commitment. Unlike the scribblings of a child, the art was a social product resulting from a strong, perhaps even urgent, impetus. As Paleolithic cultures invented themselves, they for some reason found it necessary to invest human and material wealth into the process of inscribing these images on inaccessible places. Although it may not be possible to ascertain exactly what motivated them, it is clear that they did expend resources commensurate with something of the highest importance.

These complex, durable, and hard-to-reach visual displays most likely incorporated at least some elements their creators considered significant. Of course, the art was not something meant for connoisseur viewing and the caves were not museums. Today art is divided from the routines and rituals of everyday life, but that separation likely did not exist in Paleolithic societies. One can imagine the making of such paintings as part of rites to mark some stage in a recurring cycle along the life course or the round of seasons. Archeologists suggest that prehistoric artists left a record of the exceptional or ritually important elements in their societies: symbols of fertility and hunting.

My students' initial interpretation of these symbols reveals less about Paleolithic values (which geometrical thinking would want them at least to imagine) than about today's. The myths of progress and social order, with man as hunter or provider and woman as animistic nature, are reassuring. They assert our cultural superiority over a backward past and project our system into a stable but expanding future. That mythology is then disguised by the assumption that Paleolithic art is a direct result or simple representation of life-as-lived-back-then, easily interpreted from our advanced state of affairs.

The student criticism of unfamiliar cave art reveals how the tacit acceptance of linear perspective drives expectations while limiting and narrowing them. The lack of unitary perspective in cave paintings, for example, may be not a *lack* in any real sense. The images seem to propose a collective art, cumulative and communitarian rather than individual-inspired—concepts foreign to today's definitions of art and authorship. The students' reactions grew out of the dominant formal system of today and give at least a clue about what some young people in European and American culture now consider important.

Such a system places a high value on hierarchy and reflects that value using a variety of strategies, including variations in size or scale, position or direction, and detail or texture. These modes of spatial organization are entirely conventional, and to call them *sophisticated* is a way of patting the culture of linear perspective on the back. Only ethnocentrism would propose such a basis for judging the cave paintings.

What seems ordered and logical in the U.S. mainstream today is not necessarily so for other cultures, or even for my Europeanized students. Lived experience is not without evidence of Paleolithic-style vision (figure 9). Life, or today's image of it, seems just as convoluted at certain times and places, even in that most ordered of spaces: suburban America. These days no one's work is ever solely one's own. In any multicultural society, a single scene may be viewed from complex, mobile perspectives, and events experienced without a unitary optical focus. That, I suspect, is what Waisanen was trying to project. Even if people experienced their happiest moments outside systems of hierarchy and when least aware of themselves as individuals (as in the primal tequila party), the central gaze of Western perspective reduces all things to a linear process arranged for consumption in linear pictures. This is indeed a peculiar way to deal with a complex world.

Geometrical thinking also calls into question the initial reactions to the example of ancient Roman art. The assumption that the wall painting in Pompeii is merely inept or crude blocks the imagination from seeing an alternative complexity. The contradictory perspectives in the painting manage to present a series of events simultaneously. The front elevation of the amphitheater identifies the setting. Choosing a higher, three-quarters angle for the arena and peeling back the canopy both work to reveal the event inside, while retaining the physical link between the elevation and plan views. The fact that the surrounding figures all appear at about the same scale, regardless of their locations in physical space, helps create an image of unified action. The use of outlines and silhouettes also enhances the uniformity of the depiction. Such techniques, although not the first that came to my students' minds, are clearly anything but simple. Multiple vantage points are often used today to narrate complex events. Graphic designers illustrate processes through time and across locations using time-series diagrams (figure 10). From this viewpoint, the Roman artist cannot be said to lack sophistication.

That doesn't mean, however, that Romans invented the diagramming techniques found in periodicals today; quite the contrary. In the case of visual history, the analogy of past to present can go too far, turning things like funerary art in Egypt and China and fertility art in Africa and Polynesia into the equivalent of announcements for weddings, births, and deaths in a local newspaper today. The mere presence of related kinds of information in ancient painting

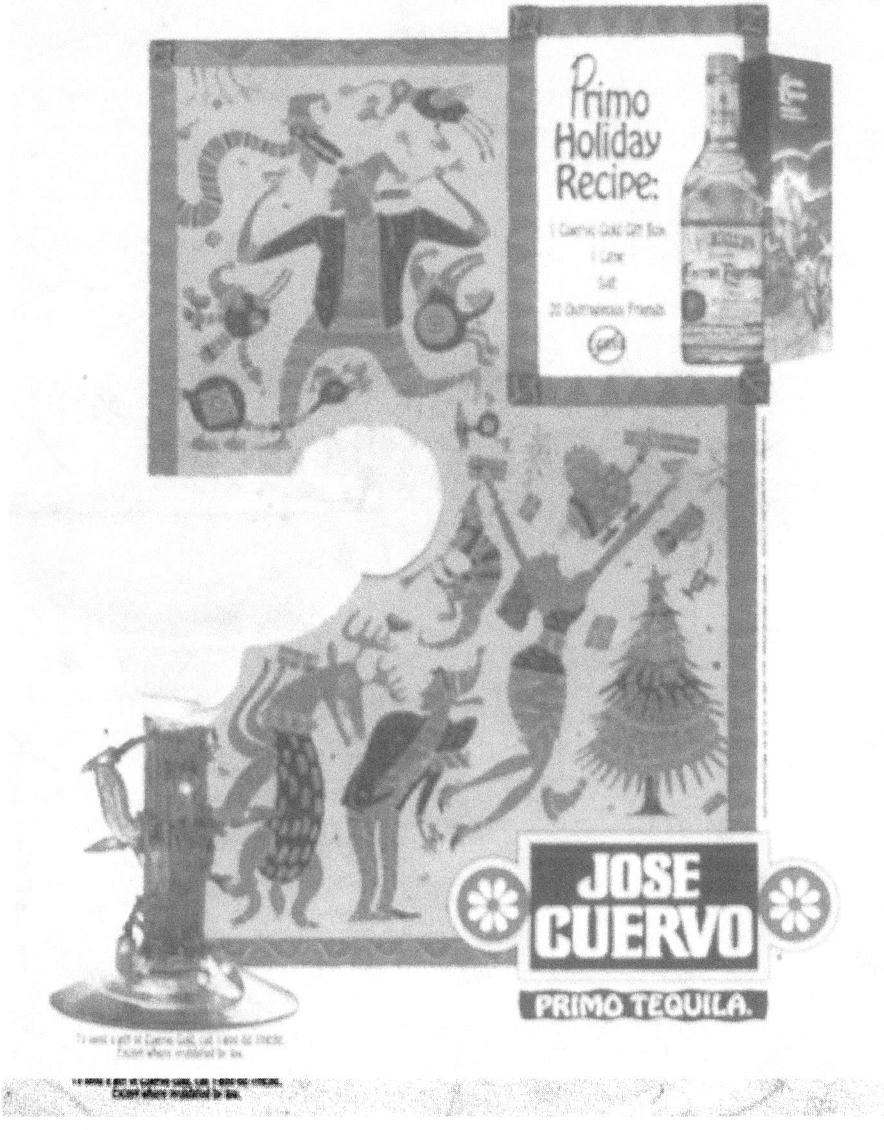

Figure 9. The holiday advertisement for José Cuervo tequila uses a primitive style: animals defying vertical and horizontal conventions, similar to prehistoric versions of perspective, with people allowed only to lean away from the vertical as an expression of the alcohol-party spirit. The perspective is carefully arranged so that the out-of-control experience is an end-product of the linear process beginning with production and moving through advertising to consumption. The zone of primitive drawings is carefully corralled by a decorative border and dominated by the consumer offer and product image (top right), the serving suggestion (left), and the brand name (bottom right).

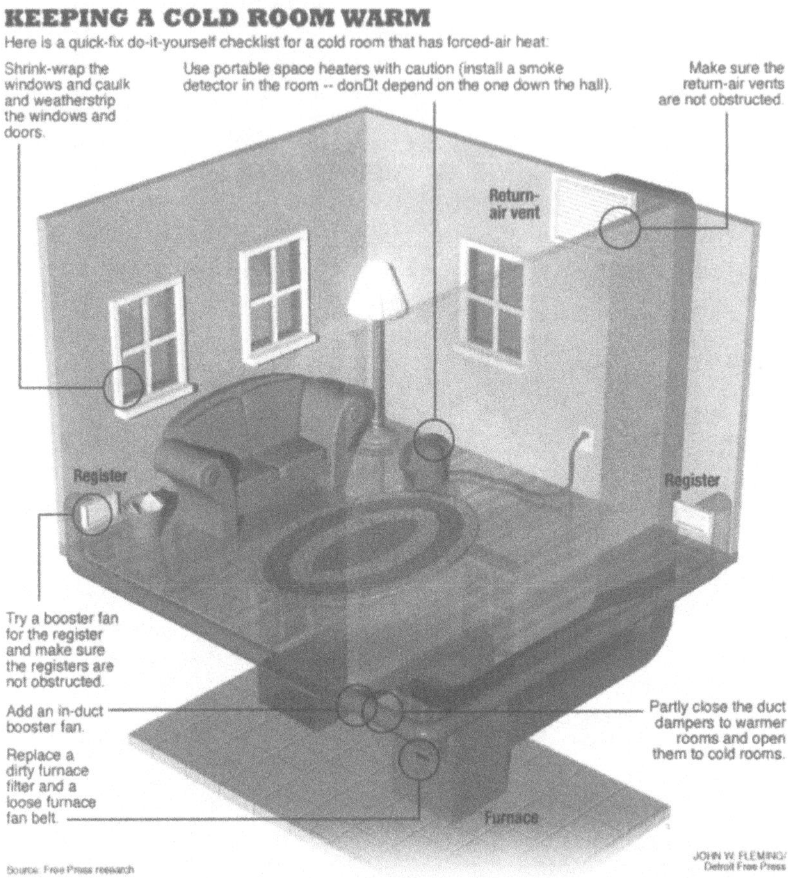

Figure 10. In this diagram, which appeared on the cover of the *Detroit Free Press* Real Estate section (Sunday, 14 January 2001:1h), the home is conceived by graphic designer John Fleming as a space surrounded by mechanical systems and devices. It is no coincidence that such a space is imagined as rectilinear and is thus rendered according to linear perspective. Note how the round throw-rug and overstuffed sofa use chiaroscuro to create the effect of coziness, so that a harsh, linear concept of private space seems completely natural.

and modern communication is inevitable. Thinking geometrically involves breaking the habit of equating the past with the present. The parallels people draw between modern media and ancient painting are evidence of the cultural values of today, not those of Rome.

Wearing a linear-perspective monocle to peer back into history makes it hard to resist the conclusion that early cultural forms somehow caused or led to the present order of things. But the assumption that art progressed is questionable at best. Although the painting in Pompeii was preserved, it is exceptional. Very few such records exist in stone and even fewer in painting. In art history, some works survive because of their durable media, others because they are valued by their immediate heirs, still others because of the wealth and power of several subsequent owners, and yet others because of dumb luck. As fewer and fewer artifacts remain, the accident of survival makes a limited number of works bear more of the explanatory weight for the culture. Newly discovered artifacts seem even more significant, but their long period out of sight limits their influence. The painting from Pompeii could not lead into modern graphic forms while it was buried for more than a thousand years. Geometrical thinking can act as a reminder that the evidence from ancient cultures is at best haphazard.

Although my students found looking for the familiar in ancient art a useful metaphorical lens, the metaphor was not passive. They tended, for example, to think of communication as the simple or direct transmission of pure information. The process of communication in Rome was much more monumental and commemorative than anything common today. To inform the citizenry of and arouse their support for military exploits, for example (Brilliant, 1974), Romans carried paintings of battles and victories in triumphal procession and then recorded some of them on the walls of buildings and monuments as friezes and wall-paintings. All metaphors actively create images, and by insisting that viewers remain conscious of what metaphor they use, geometrical thinking protects viewers from playing the smug heirs of history.

Thinking geometrically resists casting straight lines that project backward into history (or laterally into other cultures). Early societies did not dream of becoming like us because they could not have pictured us. On the other hand, we can look back and at least attempt to imagine how early societies pictured themselves. Every artifact is in the end about the culture that created it. The Columns of Trajan and Marcus Aurelius in Rome depict events, scene after scene spiraling around the stone cylinder to the top. A linear thinker would accept these as evidence of events spiraling continuously upward through history, but a geometrical thinker might see them as evidence of cultural values rather than of historical realities. Art historians have noted, for example, that the extant art of Rome seems to play self-consciously to their idea of history.

Thinking geometrically proposes that viewers seek out and recognize alternative points of view. Paleolithic cave paintings may have had no consistent horizon, but they encourage and accommodate many different viewing positions, none of them revealing the same relationships. Roman wall paintings present a pedestrian's vision of city life; that is, rays extended from the edges of buildings scatter in every direction. In some medieval paintings, all of those rays converge on a single spiritual figure. In the years after linear perspective is said to have blossomed, artists experimented with other perspective systems.

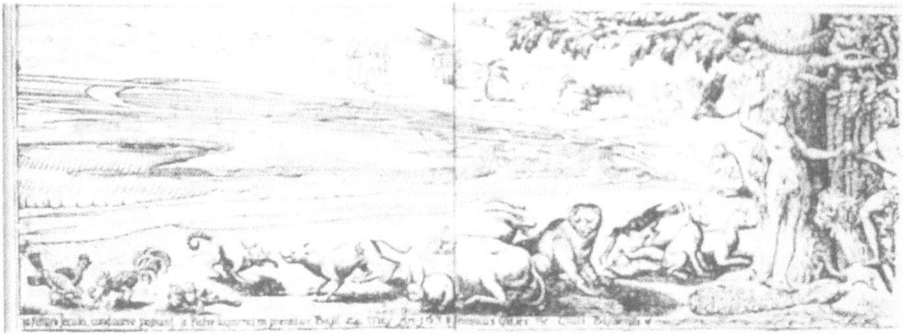

Figure 11. In the 1638 engraving *The Fall from Grace*, Johann H. Glaser shows Adam and Eve on the right beneath the tree of the knowledge of good and evil with the animals they've named playing together along a shore. The lake, when viewed from the extreme right of the frame, is another image, *Christ with the Crown of Thorns*. The engraving illustrates the dominance of linear perspective by treating the story of Adam and Eve as the main text and the redemption of Jesus as a sign of the argot decipherable only by members of the Christian subculture (see Descargues, 1976:12–13).

Some engravings from the seventeenth century used *anamorphic* perspective, with the picture plane slicing at an angle almost perpendicular to the line of sight, decentering the observer. Viewed straight on, paintings in anamorphic perspective have unintelligible passages, which can be read only diagonally from the extreme edge of the picture (figure 11). The best-known anamorphic image is found in *The Ambassadors* by Hans Holbein the Younger (1533, oil on panel, 206 by 209 cm, National Gallery, London). A large object shaped like a cigar or baguette floats in the foreground, and only the decentered viewer can see a human skull, a portent of the invisible or yet-to-come spiritual reality. Two versions of realism coexist in such paintings. In a culture saturated by the metaphor of linear perspective, the alternatives seem like strange dead ends in the progress of picture-making, when in fact they represent alternatives that open up different sets of cultural values.

Art from cultures that are more familiar, such as the nineteenth-century painting from Spain, presents other pitfalls. If viewers assume that their own culture is similar to (or growing out of) the culture that made a picture, they may look for the facts and skip the harder questions. By seeming readily decipherable, Goya's *The Third of May* deflects critical attention away from itself and into questions about its manifest content: Who is depicted? Where did the executions take place? and the like. The who, what, when, and where may be as much a barrier as an avenue to understanding. For viewers with little interest, the details are old news, just more dry facts to be memorized about history. Viewers who find the painting initially intriguing, however, may be satisfied by those same facts without going beyond them. Not knowing facts may make a picture unintelligible as content, but knowing those facts may create a false sense of understanding. These are the conditions that also make the content of newspapers and newscasts both engaging and numbing. Viewers are curious about details but feel they have heard it all before.

Studying the iconic and formal (as opposed to factual) content adds another dimension. Goya placed the firing squad facing left, contrary to the predominant reading direction in Europe. He kept the soldiers' faces hidden but showed the Madrileños confronting the rifles with their faces openly visible. In making these formal choices, he clearly was collaborating with one side of the conflict. By conferring individual identity on the resisters, Goya took a stand with them. He did not face the same risks himself; he waited until the occupying French had withdrawn to make his painting (Licht, 1979). But he did celebrate the courage of the resisters, and his painting elevates their deaths to martyrdom. The lantern is a mechanism of vision, and its light, although blocked by the soldiers, illuminates and celebrates the central human form. In contrast, Goya fused the rifles with the bodies of men in the firing squad to create a countervailing mechanism, with a resulting darkness that suggests a technological death for the executioners. These mechanisms in the painting express a profound ambiguity. Technology is the means for domination and death but also the source and reflection of faith. After a close analysis of the painting's iconic form, the argument (growing out of linear perspective) that instruments are superior seems an empty one. The hand of the artist was partisan and expressive, as opposed to the technology of the automated camera. The painting is clearly much more than a report of an event that could be accomplished through mechanical means; it is a symbolic creation.

The Third of May allows Western viewers to replay and illustrate their own culture's grand narratives. These include faith in the value of human life and in the sovereignty of the nation state, myths that are emotionally charged and come laden with references to such things as Christian atonement. The bright figure extends his arms as in the icon of Jesus on the cross, and the fallen figure before him mirrors the pose. The story of crucifixion involves the spiritual independence of individuals and also their subordination to government. *The Third of May* on one level is about the life after death and the resurrection of the just Spanish state, but at another level it is about the enshrinement of the self. The highest honor in linear perspective goes to the individual, whose fate is central to the survival of governments and to the meaning of life.

Goya's selection of icons and his arrangement of them in visual structures that invoke grand narratives has been frequently imitated, whether consciously or not. Several recent advertisements on U.S. television have shown a central figure with outstretched arms, sometimes isolated beside a powerful automobile atop a towering rock or mesa. Instead of shining a brilliant lantern, the ad focuses attention by circling around the lone figure, and danger (or death) is left implicit by the steep drop-offs in every direction. The viewer is invited to ask questions of fact—Who is this person? How did he get there?—but not to question the myths of technological power or individual autonomy. Thinking geometrically demands that other dimensions get considered, including among other things the surrounding history, the formal structure of the image, its iconography, and the mythology it invokes.

Beyond the Frame

No perspective replicates precisely what humans see or what is out there physically. Perspective systems are like precision instruments; their measurements provide sharp, clear insights, but not without limitations. Hierarchy in a newspaper page makes editorial priorities immediately clear, but that strength is also a weakness. Some people or ideas get persistently pushed to the bottom of the layout, toward the back of pictures, or out of sight entirely. Readers who disagree with the editor's priorities may resist the system by starting elsewhere, but they are more likely to abandon newspapers altogether. The same is true for any medium of communication. The world cannot be neatly prioritized without removing conflicting points of view. Internet sites that are meant to attract and direct the viewer's eye turn hierarchy into a tool of marketing. Media that accommodate competing priorities may be more egalitarian, but not without making demands on their audiences, who must sort things out for themselves.

The illusion of realism that serves to make the visual hierarchy authoritative cannot be easily turned to other purposes. Simply switching the panoramic quality of conventional perspective to make it a backdrop for images of individuals who lack power usually backfires. A grand vista can take on an ironic significance, crushing or looming over anyone poor or exotic-looking. On the other hand, surreal effects applied to, say, a head of state can appear to be just another politically motivated attack. Although the pattern of reinforcing hierarchy with realistic illusion is hard to break, media workers are still responsible for the choices they make. For example, rectifying the horizons of battle pictures removes evidence of the chaotic experiences of a photographer among combatants. Following the standard procedure makes war look good—controlled in some way, or natural—and even the most visually astute viewer cannot detect the alteration. Geometrical thinkers, who understand systems of perspective, would not introduce such distortions, but the resulting images might seem more random, sacrificing the polish of professionalism.

To recognize the limitations of any perspective requires thought and study but also depends on the ability to compare the illusionism with one's own eyesight. In the case of chiaroscuro, careful examination of objects such as a ball and a shoebox under a light will reveal the limitation. If they in fact occupy opposite sides of either of those objects, the highlight and deep shadow could hardly be visible at the same time. Chiaroscuro distorts what the human eyes see, in order to create its illusion. Atmospheric perspective also yields to observation. Anyone traveling across country can note how often the more-distant hills or mountains seem darker and larger rather than fainter. In both cases, critical inspection exposes ways that the conventional systems differ from eyesight. These are the sorts of observations that may act as the test Waisanen suggests for any system of visual thinking.

Linear perspective, which he discusses in depth, has limitations that are likewise easily demonstrated. When people stroll down a long sidewalk, or better still, along some railroad tracks crossing a flatland or plain, they may imagine themselves in a classic painting or photograph, walking within a linear perspective world. In the distance, the tracks or sidewalk edges converge toward a visible vanishing point. But looking down, they will see, as they stand between those same tracks or edges, that the lines are parallel. Although rarely attempted, it is possible to position one's eyes somewhere in the middle, relax the focus, and take in the lines as they curve across the visual field. From the distant convergence, they open up gradually and move past, like water opening for a boat prow and swirling in its wake. The experience of

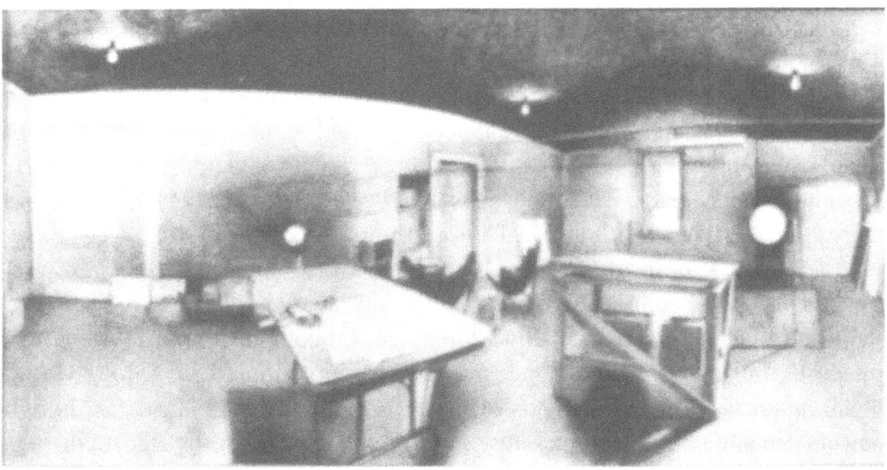

Figure 12. William Ramage, *Studio Drawing* (graphite, 24.384 meters), from "Works of Art," an exhibit at the Brattleboro (Vermont) Museum of Art in 1984.

vision, as people move along the structure of roads and corridors of modern life, might easily be conceived as curvilinear, although people resist the idea.

Twentieth-century artists developed several systems that record the curvature of the visual field. M. C. Escher, perhaps the best known, did convincing studies of curved perspective in the 1950s, and Albert Flocon produced an exacting system in engravings of the 1960s. One summer in New England, I happened to visit an art gallery in Brattleboro, Vermont, where works by William Ramage were on display. From most vantage points, his *Studio Drawing* may look strange (figure 12), but viewed from a spot about a yard away (marked clearly on the gallery floor), everything looked perfect. The perspective suddenly seemed natural, as if seen from inside the depicted studio itself. (Of course, the position marked on the floor was where most viewers could easily adjust the image to their expectations of linear perspective.) Curved perspective builds from a complex geometry of arcs, but nongeometric systems are also possible.

By seeking alternatives, Waisanen has proposed a way to expand the imagination. He criticizes visual conceptions that are "concerned only with . . . imitative or realistic representation" (75). The idea that imagery can look like an external, real world is sometimes called *mimesis*, and Waisanen initially invokes mimesis as a justification for geometrical thinking. Later his argument recognizes the need to break free from the constraints mimesis has imposed, at least through the widespread use of linear perspective, the commonly accepted mimetic norm. Although he writes frequently of nature, he is clearly making a case for shifting visual thinking toward culture, for training the technologies of vision upon social space. Thinking geometrically makes visible the complex relationship between the social world and mimesis as a mechanism of realistic representation.

Rebuilding the mechanisms of imagination through geometrical thinking would likely cause discomfort. The experiments of Helmholtz in the nineteenth century suggest that the human mind would be capable of making sense of many possible visual systems. Although his subjects experienced intense nausea and confusion when fitted with unusual lenses, they soon adjusted. For a week or so, they saw the world upside down and yet grew accustomed to eating things on the (upside-down) table, sleeping on the (upside-down) bed, and so forth. Eventually nothing seemed upside down anymore. Of course, when they removed the lenses they had to readjust once more, just as violently, to the regular world, which again was turned upside down without the unusual lenses. Although viewers can adjust to different ways of seeing, Web pages and other media images arranged by alternative systems would not be easy at first to parse. By decentering the viewer, however, they would allow for other philosophical positions that start from diversity or community instead of from the individual.

The first to adopt geometrical thinking would be the visually literate: the designers of Web sites, the creators of advertising, the photographers and layout artists at newspapers, and others involved in applied art, as well as fine artists and visual scholars. Painting, for good or ill, is the most studied visual medium, and for that reason alone it deserves attention first. In this initial foray into geometrical thinking, space has not permitted a detailed examination of the political or economic situation surrounding the images, let alone the broader conditions to the extent they are known about the cultures, although that is what Waisanen has called for. As it turns out, there is a project that does something along those lines, which I will describe in closing.

In a book-length analysis of form through U.S. history, John Nerone and I (Barnhurst & Nerone, 2001) had the space to examine images of newspapers fully contextualized within patterns of narrative, shifts in predominant visual styles, conditions of production, and wider cultural changes. We did not know of Waisanen's work, but our research for each major era of U.S. history found something akin to the results he anticipated from the process of thinking geometrically. By taking each of the perspectives into account—the macrolevel political and economic dimensions as well as the microlevel stylistic and narrative dimensions—we discovered an image of U.S. political and cultural life projected through the forms of news. In each case, the image could best be expressed as a metaphor, the product that Waisanen said would result from geometrical thinking.

To summarize our results briefly, the metaphor that reigned from the colonial era through the early American republic was a *town meeting* among citizens, who were enfranchised by their position as propertied gentlemen. They would meet in coffeehouses and in the columns of early newspapers, which acted throughout the printerly era as extensions of public deliberation, resulting in the forms of news associated with artisanal printing. After large shifts in the cultural landscape of the United States, including the rise of the market economy and the advance of democratic ideals associated with the Jacksonian era, the town meeting metaphor no longer fit the prevailing image of public life. As these perspectives changed, so did the conditions of print production, accompanied by a stylistic evolution that played out in the appearance of newspapers. As a consequence, the initial metaphor was slowly replaced. By the 1820s the country had entered the partisan era, and public life as presented in the newspaper had the qualities of a *courtroom*, in which competing interests, economic as well as political, vied for power. The further advance of the market, along with the growth of cities, transportation, and communication, was accompanied in the Victorian era by technological changes in print production. These currents in the culture led to a new metaphor, the *marketplace*, and newspapers became compendia of items for purchase and consumption.

The modern era for newspapers emerged in the twentieth century, along with the further development of print production and the professionalization of news work. The modern metaphor, a social *map*, allowed the newly endowed professionals of journalism to order and prioritize the stuff of public life for an audience of otherwise uninformed and passive citizens. Then, as corporate modes of organization came to dominate U.S. society and the production of news as well, the metaphor shifted again. In the latest phase of the modern era, the controlling metaphor has become a multidimensional *index*. As it evolved along with other visual arts and moved onto computer networks and spread internationally, the index metaphor also found expression through the form of news. Any index hides its organizing authority by pointing elsewhere. An Internet link makes a Web page richer by adding connections, and news in print uses a related vocabulary attributing content and referring readers to other locations and articles within its pages and to other sources beyond its pages.

In our version of the theoretical move Waisanen has taken, we examined what he might consider geometric projections, and from them a picture emerged of the reigning metaphor for each era. We imagined each metaphor as a multidimensional space, which readers were invited to enter. That space forms a media environment. The image of persons and events it projects is not objective or neutral but ideological. When readers choose to occupy the newspaper environment, they are hailed into a network of relationships, both material and symbolic. They are citizens of real countries, consumers of real products, and fans of real teams, but they are also readers given less say in any matter than journalists have, subjected to limited choices in the market, and reduced to passivity as spectators who are powerless to do more than express emotion. Whether wittingly or not, they are positioned in a network of relations not only with the concrete newspaper and its advertisers but also with symbolic representations of politicians and celebrities, institutions and other entities. The newspaper environment opens up a particular vista of that network while it also closes off other potential vistas.

Theorists have conceived of the news media as either transmitting messages to receivers or sustaining the existence of communities through rituals. But messages are not simple reflections or bits of the world-out-there, and communities do not simply use media to perpetuate themselves. We argue that messages and communities are involved in another, represented, order of relationships. Journalists, politicians, advertisers, readers, and anyone else involved use the newspaper to imagine themselves in relation to everyone else, and together they create an artifact that expresses not what is but what *is imagined*. The form of news, while allowing the messages to change from day to day, remains constant enough to reiterate and sustain community by establishing a metaphor for that imagined world. Our notion of form is built on the *represented* order, which is perhaps just another phrase to describe the projections required for thinking geometrically.

The Internet is the newest form of represented relationships. Today everyone, it seems, must have a Web site, and that site must show (or imagine for others) some aspect of one's life. Everyone is also expected to venture out into other sites, to encounter other imagined lives. The need for geometrical thinking is especially urgent in these zones of computer-based communication, where the visual text is constantly present and where intercultural encounters occur with greater frequency than perhaps anywhere else. The Internet is of course not part of everyone's concrete, hands-on experience, but everyone has direct exposure to the Internet as a metaphor. The Internet is itself an imagined environment. It expresses how the mainstream in the United States and elsewhere is coming to recognize the need to reach across cultural divides, rethinking cultural assumptions. In that context, thinking geometrically can have the effect of extending multicultural thought.

Waisanen has proposed not a different mathematical formula for seeing the world but a different orientation to knowledge itself. He discredits that old saw, seeing is believing. Instead of leaping thoughtlessly from a habitual vision of the world into the dead end of unquestioned knowing, he proposes a pause, a moment for hedging and for making perhaps better, less overdetermined choices. He is right to question the dominant, linear perspective, and right to leave with us the charge to look for alternatives now, at the Internet frontier. More than a century ago another technology of communication, the railroads, built a network across the Great Plains in the American Midwest. That grid and the surveys and county roads that came with it threatened and then destroyed the landscape as known to its original occupants. In his eloquent history of the region that extends "beyond the frame of the photograph," Ian Frazier opens up a small gap in the pervasive linear curtain: "Places Indians liked are different from what you see while driving the grid of roads stencilled over the Great Plains. Sometimes when you find such a place it makes that grid seem to disappear" (1989:214, 38).

NOTES

Chapter 2. Thinking Geometrically

1. Geometrical proofs that depend on graphic demonstration are today considered suspect (Kramer, 1970; Pedoe, 1976).

2. Originally published in Spanish in 1956 as *En Torno a Galileo,* literally, "About [or Around] Galileo." The English translation, *Man and Crisis,* wasn't published until 1958.

3. The importance of Galileo's theoretical work tends to overshadow his practical accomplishments, which have had an equally far-reaching effect on the development of Western technological culture. For example, he devised major improvements to the optics of the telescope, which he was the first to use for astronomical observation, and he suggested the use of pendulums for the regulation of clocks. Without these contributions, modern standards of experimental rigor and precision of measurement would be impossible.

4. He also demonstrated that imaginative work doesn't necessarily find a convenient place in reality. The last eight years of his life were spent under house arrest for contradicting Church dogma.

5. Analytic geometry, the mathematics of this coordinate space, unites geometry and algebra and provides the scientific basis for perspective.

6. An interesting byproduct of the ascendance of mathematics after Galileo is that logic is now perhaps more attached to mathematics than it is attached to language.

7. Gardner's (1983/1993, 1993) seven intelligences are linguistic, musical, logical-mathematical, spatial, bodily-kinesthetic, and two personal intelligences: access to one's own feeling life, and the ability to notice and make distinctions among other individuals. Gardner's concept of spatial intelligence is more limited than the concept of space I am arguing for. He defines spatial intelligence as "capacities to perceive the visual world accurately, to perform transformations and modifications upon one's initial perceptions," and the ability "to re-create aspects of one's visual experience, even in the absence of relevant physical stimuli" (1983/1993:173).

8. *Rashomon* is a classic film directed by Akira Kurosawa (1950) that depicts four different versions of the "same" rape and murder. The film has become the archetype for many narratives since that explore the idea that multiple points of view can render reaching the "truth" impossible.

9. Even Newton's concept of gravitation, which may appear to explain why bodies fall to earth, cannot itself be explained. We still don't know what gravity really is or what causes it.

10. Early researchers into electrical phenomena were stymied by their reliance on mechanical models; Maxwell's equations, the basis for a science of electronics, have no visual or mechanical analogue. They can only be understood mathematically.

Chapter 3. Stories of Space

1. "Parabolic" here carries the meaning of "like a parable" rather than its geometrical connotation.
2. This point was taken from Havelock's *Preface to Plato* (1963).
3. The Incas, for example, managed for centuries without a written language.
4. For examples of explorations of multiple dimensions in art and science, see Robbin, *Fourfield: Computers, Art and the 4th Dimension* (1992); in mathematics and geometry, see Rucker, *Geometry, Relativity and the Fourth Dimension* (1977); in physics, see Gregory, *Inventing Reality: Physics as Language* (1998).
5. Although Rucker and Robbin both refer to the Author as "A. Square" there is no explicit reference as such in Abbott's book. In the preface to the second edition, Abbott refers to "A. Square" but everywhere else, he is just the "Author" and I will emulate this usage.
6. As Abbott explains, the fact that they see each other at all implies a certain degree of actual thickness (a line without thickness would be invisible!) and thus three-dimensionality, but this is essentially imperceptible to the inhabitants of Flatland.
7. In this regard, Rucker (1977) points out a fundamental flaw in the story of the Author's voyage into the third dimension: Since his flat eye(s) would have been capable of only a single line of vision, he would have been physically unable to perceive the third dimension except as a series of two-dimensional perceptions. This objection is fairly successfully dealt with by Abbott in the preface to the second edition of *Flatland.*
8. This axis always points toward the center of the earth, and its direction therefore varies according to where you stand on the surface of the globe.
9. Intermediation, taken to an extreme, could mean piling on level after level of mediation. Without being careful, the philosopher might well end up, like the poet or artist, at several removes from true knowledge.

Chapter 4. Drawing Space

1. The very irregularity of ancient city layouts may have also inhibited the development of perspective. The multiplicity of vanishing points produced by buildings that are not square to their neighbors makes the systematic convergence of perspective extremely difficult to recognize. In contrast, the single vanishing point of a view down a long avenue of parallel buildings makes this convergence obvious to the eye.
2. Comte (in Hall, 1902/1903:7) calls attention to this division in his acknowledgement of the "high industrial utility" of descriptive geometry. He does not go so far as to claim that it is equally useful to philosophy; rather, he says that it has a "philosophical *peculiarity.*" Even as he extolled its advantages to the imagination, he diminished its general importance through his choice of words.
3. The Romans also invented concrete, which simplifies some aspects of construction but introduces other problems. For example, although stonecutting is nearly eliminated, the need to design and build accurate forms and scaffolding is introduced.

4. Even so, many empirically developed practices established well before the invention of descriptive geometry survive in various branches of drafting, such as architectural drawing.

5. As a required subject in most engineering programs, descriptive geometry abounds in texts. However, even after 200 years of development, no single clear and concise account of descriptive geometry yet exists. Descriptive geometry is a difficult subject to master on its own terms, and even harder to put into words. All the texts I have seen on the subject are tedious and complex in the extreme. As long as this is the case, descriptive geometry—like most other technical subject matter—remains a door closed to the general public, and the key role it has played in the production of contemporary space remains unrecognized.

6. Fortunately, there are a few good and relatively accessible discussions of the significance of orthographic engineering drawing: Eugene Ferguson's *Engineering and The Mind's Eye* (1992) documents the important role of nonverbal learning and thought in the development of engineering generally and in engineering's development of the spatial/visual imagination specifically. Peter Booker's *A History of Engineering Drawing* (1979), a slender volume in second edition but now out of print and rather difficult to find, develops a slightly different direction that I have drawn on extensively in this chapter.

7. Writing, strictly speaking, is not a representation of spoken language—letters and words do not in any sense "look like" their spoken counterparts. Mathematics is pure abstraction—the power of numbers rests in that they stand for concepts, quantities, and relations, and not objects.

8. Projective geometry would not become fully developed until several centuries later.

9. Monge ought to be remembered for being as committed to human concerns as he was to the advancement of theoretical knowledge. He was fully involved, ultimately to his own detriment, in the social, economic, and political issues of his day. Contemporary engineers, often accused of a lack of concern for the social, environmental, and cultural effects of their work, could do worse than to follow his example of political commitment and social involvement. Perhaps the abrupt and unhappy ending of his career and life serves as a cautionary parable of greater impact than any of his earlier accomplishments.

10. Interestingly, timekeeping concerns converge with those of shipbuilding in mastering the problems of early transoceanic navigation, in which the accurate determination of longitude is not possible without some kind of reliable chronometer, which was not invented until the eighteenth century (see Sobel, 1995).

11. See Pedoe (1976) for an interesting account of conics.

12. The conventional wisdom in design professions is that it is cheaper to make (and correct) mistakes on paper than it is in the building.

13. None of Monge's work or biographies as yet exist in English translation, which doesn't help this cause. Further, Ortega y Gasset's (1958) concept of the role of the "generation"—roughly, 15 years—in cultural change is here illuminating. The delayed publication of Monge's work for 30 years—two generations—may have put his work past its optimal "window" in time. In this duration, the world had already moved into a new phase of modernity.

14. Secondary geometries, though usually guided by sound theory, are no more than localized and compartmentalized applications and limit the range of a practioner's knowledge.

15. In drafting, a different terminology is commonly used; the plane of projection is labeled simply by the missing axis. Thus, the plan view is located in a Y plane, the front view is located in a Z plane, and the side view is in a X plane.

16. For example, in automotive design, the "side view:" is functionally equivalent to the "front elevation" in architecture.

17. This view is technically an "isometric" projection, which is discussed in more detail below.

18. One drawback to the third quadrant, in strict Cartesian terms, is that its coordinates are all negative. In common practice, however, positive coordinates are used instead for convenience.

19. Anamorphic projection, a drawing and painting practice developed in the late Renaissance, insists on confronting the (potential) difference between the location and angle of the artist and the viewer. In anamorphosis, images appear distorted when viewed in front view, but when the viewer adjusts their position to the side and/or above or below to look over at the picture plane from an oblique rather than 90% angle, the image approximates what would be expected from a front view. Thus the location and angle required to "see" the image does not mirror the location and angle of the artist in relation to construction of the image.

20. This image circulates so widely, it is often not even credited. For example, it's on the cover of Lynch and Woolgar's *Representation in Scientific Practice* (1990) without attribution.

21. This kind of thinking is also demonstrated by air traffic controllers, who must construct a three-dimensional model of objects in space from the two-dimensional graphic evidence of a radar display.

22. Comte, who coined the disciplinary label "sociology," was a positivist and a staunch believer in the power of science and logic. "Positive," as he uses it, does not carry the contemporary meaning as the antonym of negative. By its "acceptation," one becomes a believer in the power and rightness of logic and science, a position which I do not entirely accept.

23. The growth of technical communication programs is no doubt partly due to the widely recognized difficulty many science and engineering students experience in communicating their concepts to a wider audience in either spoken or written plain English. It is not my intention here to explore the question of why this native language proficiency is not necessarily carried over into scientific and technical practice. At this time I can only acknowledge the problem and speculate that it may be one based as much in the inadequacies of language as a medium for spatial thinking as it is in the absence of humanities courses in scientific and technical curricula.

Chapter 5. Designing Space

1. These connections (or interrelations) are roughly parallel to Lefebvre's three "moments" of space: practices of space, representations of space, and spaces of representation. They also serve as determinants of the current "mode" of space.

2. In this, it is no different from "science," which is similarly spread among multiple disciplines and directed at multiple concerns. This multiplicity is amended through the nearly universal acceptance of the "scientific method" among these disciplines. There is as yet no such "design method" to connect the work of designers in different fields.

3. Arriving at an unambiguous definition of the difference between text and graphics is most difficult. For example, line drawing, perhaps the most basic form of drawing (as in stick figures) is produced with the same instruments, materials, and techniques as the most basic form of true writing (phonetic script, the letter forms of which are traceable to pictographic origins); both stick figures and letters are equally abstract representations of their subject matter (objects and sounds, respectively). Terms like "orthographic" apply equally to systems of graphic projection and writing projection; the processes and products of

"composition" are equally graphic and textual in character. Text and graphics are inextricably interwoven in theory and practice.

4. It is interesting that he specifies *changes* in procedures, and not just the procedures themselves, as making a difference. This is an important point to which I will return later.

5. Mathematics might also be considered optional, since it is possible to live without all but the most basic knowledge of it without extreme social sanction. People frequently claim to have no head for numbers, just as they claim they can't draw a straight line. But it is rare that someone admits to being unable to handle language.

6. Our perceptual and conceptual equipment seems to be geared for dealing with no more than two or three dimensions at a time, which might be called "sections" or "projections" of greater dimensionality into a frame of reference of lesser dimensionality. What Gardner (1983/1993, 1993) calls "multiple intelligences" may represent different subsets of this larger space. There is no absolute reference, no center to a space of even two dimensions; different spaces may interpenetrate one another, but it is difficult to project them onto one another. As seen before in the examples of *Flatland* and Plato's Cave, projections or sections of one space into another—like the passage of the Sphere through the Author's two-dimensional space—can never be more than partial and even distorted representations of the objects or dimensions of the other. Orthographic drawing provides a systematic means for doing this in regard to graphic representations, but we have no such means of relating concepts that are unrepresentable as lines on paper. Geometry is the science of organizing the multiplicity of space.

REFERENCES

Preface References

Bowker, G. C. & Star, S. L. (1999). *Sorting things out: Classification and its consequences*. Cambridge, MA: MIT Press.

Chicago Public Radio. Loyola Cuts Classics. In broadcast of National Public Radio's *All Things Considered* (March 23, 2001).

Cooley, M. (1980). *Architect or bee? The human/technology relationship*. Boston: South End Press.

Deleuze, G. & Guattari, F. (1987). *A thousand plateaus: Capitalism and schizophrenia* (B. Massumi, Trans.). Minneapolis: University of Minnesota Press.

Gates, B. (1999). *Business @ the speed of thought: Using a digital nervous system*. New York: Warner Books.

Lefebvre, H. (1991). *The production of space* (D. Nicholson-Smith, Trans.). Cambridge, MA: Blackwell Publishers.

Merton, T. (1989). *Conjectures of a guilty bystander*. New York: Doubleday.

Nhat Hanh, T. (1991). *Peace is every step: The path of mindfulness in everyday life*. New York: Bantam Books.

Ong, W. J. (1967). *The presence of the word: Some prolegomena for cultural and religious history*. New Haven: Yale University Press.

Wilson, F. R. (1999). *The hand: How its use shapes the brain, language, and human culture*. New York: Vintage Books.

Text References

Abbott, E. A. (1952). *Flatland: A romance in many directions* (2nd ed.). New York: Dover. Originally published in 1884.

Aubry, P. V. (1954). *Monge: Le savant ami de Napoléon Bonaparte, 1746–1818*. Paris: Gauthier-Villars.

Bastide, F. (1990). The iconography of scientific texts: Principles of analysis. In Lynch & Woolgar (Eds.), *Representation in scientific practice* (pp. 187–230). Cambridge, MA: MIT Press.

Belofsky, H. (1991). Engineering drawing—A universal language in two dialects. *Technology and Culture* 32:1, 23–46.

Bolter, J. D. (1984). *Turing's man: Western culture in the computer age*. Chapel Hill: University of North Carolina Press.

Bolter, J. D. (1991). *Writing space: The computer, hypertext, and the history of writing*. Hillsdale, NJ: Lawrence Erlbaum Associates.

Booker, P. J. (1979). *A history of engineering drawing*. London: Northgate Publishing.

Buchanan, R. (1995). Rhetoric, humanism and design. In R. Buchanan & V. Margolin (Eds.), *Discovering design: Explorations in design studies* (pp. 23–66). Chicago: University of Chicago Press.

Buchanan, R., & Margolin, V. (Eds). (1995). *Discovering design: Explorations in design studies*. Chicago: University of Chicago Press.

Cornford, F. M. (1981/1982). Translator's notes to *The Republic of Plato*. New York: Oxford University Press.

Damisch, H. (1994). *The origin of perspective* (J. Goodman, Trans.). Cambridge, MA: MIT Press.

Davies, D., Bathurst, D., & Bathurst, R. (1990). *The telling image: The changing balance between words and pictures in a technological age*. Oxford, England: Oxford University Press.

Dohrn-Van Rossum, G. (1996). *History of the hour: Clocks and modern temporal orders* (T. Dunlap, Trans.). Chicago: University of Chicago Press.

Eisenstein, E. L. (1979). *The printing press as an agent of change: Communications and cultural transformations in early modern Europe*. Cambridge, England: Cambridge University Press.

Elkins, J. (1999). *What painting is: How to think about oil painting using the language of alchemy*. New York: Routledge.

Encyclopedia Britannica (Vol. 7). (1967). Descriptive geometry. Chicago: Encyclopedia Britannica.

Encyclopedia Britannica (Vol. 9). (1967). Galileo Galilei. Chicago: Encyclopedia Britannica.

Feenberg, A. (1991). *Critical theory of technology*. New York: Oxford University Press.

Ferguson, E. S. (1992). *Engineering and the mind's eye*. Cambridge, MA: MIT Press.

Foucault, M. (1973). *The order of things: An archaeology of the human sciences*. New York: Vintage Books.

Foucault, M. (1979). *Discipline and punish: The birth of the prison* (A. Sheridan, Trans.). New York: Vintage Books.

Gardner, H. (1983/1993). *Frames of mind: The theory of multiple intelligences*. New York: Basic Books.

Gardner, H. (1993). *Multiple intelligences: The theory in practice*. New York: Basic Books.

Gombrich, E. H. (1980). Standards of truth: The arrested image and the moving eye. In W. J. T. Mitchell (Ed.), *The language of images* (pp. 181–217). Chicago: University of Chicago Press.

Gregory, B. (1988). *Inventing reality: Physics as language*. New York: John Wiley & Sons.

Hall, W. S. (1902/1903). *Descriptive geometry: With numerous problems and practical applications*. New York: Van Nostrand.

Havelock, E. A. (1963). *Preface to Plato*. Cambridge, MA: Harvard University Press.

Ivins, W. M. (1946). *Art and geometry: A study in space intuitions*. Cambridge, MA: Harvard University Press.

Ivins, W. M. (1953). *Prints and visual communication*. Cambridge, MA: Harvard University Press.

Kline, M. (1985). *Mathematics and the search for knowledge*. New York: Oxford University Press

Kosslyn, S. & Koenig, O. (1992). *Wet mind: The new cognitive neuroscience*. New York: Maxwell MacMillan International.

Kramer, E. E. (1970). *The nature and growth of modern mathematics* (Vols. 1–2). New York: Hawthorn Books.

Kurosawa, A. (1950). *Rashomon*. Produced by Jingo Minoura.

Latour, B. (1990). Drawing things together. In Lynch & Woolgar (Eds.), *Representation in scientific practice* (pp. 19–68). Cambridge, MA: MIT Press.

Lefebvre, H. (1991). *The production of space* (D. Nicholson-Smith, Trans.). Cambridge, MA: Blackwell Publishers.

Lindsay, A. D. (1957). Translator's notes to *The Republic of Plato*. New York: E. P. Dutton.

Ludlum, R. (1980). *The Bourne identity*. New York: Richard Marek Publishers.

Ludlum, R. (1986). *The Bourne supremacy*. New York: Random House.

Ludlum, R. (1990). *The Bourne ultimatum*. New York: Random House.

Lynch, M. & Woolgar, S. (Eds.) (1990). *Representation in scientific practice*. Cambridge, MA: MIT Press.

McLuhan, M. (1962). *The Gutenberg galaxy: The making of typographic man*. Toronto: University of Toronto Press.

Messaris, P. (1993). Visual "literacy": A theoretical synthesis. *Communication Theory* 3:4 (November), 277–293.

Mitchell, W. J. T. (1994). *Picture theory: Essays on verbal and visual representation*. Chicago: University of Chicago Press.

Ong, W. J. (1967). *The presence of the word: Some prolegomena for cultural and religious history*. New Haven: Yale University Press.

Ong, W. J. (1982). *Orality and literacy*. London and New York: Methuen.

Ortega y Gasset, J. (1958). *Man and crisis* (M. Adams, Trans.). New York: W. W. Norton.

Panofsky, E. (1991). *Perspective as symbolic form* (C. S. Wood, Trans.). New York: Zone Books.

Pedoe, D. (1976). *Geometry and the visual arts*. Mineola, NY: Dover Publications.

Plato (1957). *The Republic of Plato*. (A. D. Lindsay, Trans.) New York: E. P. Dutton.

Plato (1981/1982). *The Republic of Plato*. (F. M. Cornford, Trans.) New York: Oxford University Press.

Poster, M. (1990). *The mode of information: Poststructuralism and social context*. Chicago: University of Chicago Press.

Robbin, T. (1992). *Fourfield: Computers, art, & the 4th dimension*. Boston: Little, Brown.

Rucker, R. v. B. (1977). *Geometry, relativity and the fourth dimension*. New York: Dover Publications.

Sacks, O. (1995). *An anthropologist on Mars: Seven paradoxical tales*. New York: Alfred A. Knopf.

Sanders, B. (1994). *A is for ox: The collapse of literacy and the rise of violence in an electronic age*. New York: Vintage Books.

Simpson, M. (1976). *Gods and heroes of the Greeks: The library of Apollodorus*. Amherst, MA: University of Massachusetts Press.

Snyder, J. (1980). Picturing Vision. In W. J. T. Mitchell (Ed.), *The language of images* (pp. 219–246). Chicago: University of Chicago Press.

Sobel, D. (1995). *Longitude: The true story of a lone genius who solved the greatest scientific problem of his time*. New York: Walker.

Soja, E. W. (1989). *Postmodern geographies: The reassertion of space in critical social theory*. New York: Verso.

Stafford, B. M. (1994). *Artful science: Enlightenment, entertainment, and the eclipse of visual education*. Cambridge, MA: MIT Press.
Turner, M. (1996). *The literary mind*. New York: Oxford University Press.
Webster's new twentieth century dictionary (2nd ed.) (1983). New York: Simon & Schuster.
Winner, L. (1986). *The whale and the reactor: A search for limits in an age of high technology*. Chicago: University of Chicago Press.
Wood, D. (1992). *The power of maps*. New York: Guilford Press.
Yates, F. (1966). *The art of memory*. Chicago: University of Chicago Press.
Zaccai, G. (1995). Art and technology: Aesthetics redefined. In R. Buchanan & V. Margolin (Eds.), *Discovering design: Explorations in design studies* (pp. 3–12). Chicago: University of Chicago Press.

Postscript References

Bandinelli, R. B. (1970). *Rome: The center of power, 500 B.C. to A.D. 200* (P. Green, Trans.). New York: G. Braziller.
Barnhurst, K. G. (1994). *Seeing the newspaper*. New York: St. Martin's Press.
Barnhurst, K. G., & Nerone, J. (2001). *The form of news, a history*. New York: Guilford.
Brilliant, R. (1974). *Roman art from the Republic to Constantine*. London: Phaidon.
Bunim, M. S. (1940). *Space in medieval painting and the forerunners of perspective*. New York: Columbia University Press.
Descargues, P. (1976). *Perspective*. New York: Harry N. Abrams.
Edgerton, S. (1975). *The Renaissance rediscovery of linear perspective*. New York: Harper & Row.
Frazier, I. (1989). *Great plains*. New York: Farrar, Straus, & Giroux.
Lambert, S. (1984). *Drawing: Technique and purpose*. London: Trevoil Books.
Leroi-Gourhan, A. (1982). *The dawn of European art: An introduction to Paleolithic cave painting*. Cambridge, MA: Cambridge University Press.
Licht, F. (1979). *Goya: The origins of the modern temper in art*. New York: Universe Books.
Panofsky, E. (1991). *Perspective as symbolic form* (C. S. Wood, Trans.). New York: Zone Books.
Schafer, H. (1974). *Principles of Egyptian art* (E. Brunner-Traut, Ed., J. Baines, Trans.). Oxford, England: Clarendon Press.
Sieveking, A. (1979). *The cave artists*. London: Thames & Hudson.
White, J. (1987). *The birth and rebirth of pictorial space*. Cambridge, MA: Belknap Press.

INDEX

Digital Formations

General Editor: Steve Jones

Digital Formations is the new source for critical, well-written books about digital technologies and modern life. Books in this series will break new ground by emphasizing multiple methodological and theoretical approaches to deeply probe the formation and reformation of lived experience as it is refracted through digital interaction. Each volume in *Digital Formations* will push forward our understanding of the intersections—and corresponding implications—between the digital technologies and everyday life. This series will examine broad issues in realms such as digital culture, electronic commerce, law, politics and governance, gender, the Internet, race, art, health and medicine, and education. The series will emphasize critical studies in the context of emergent and existing digital technologies.

For additional information about this series or for the submission of manuscripts, please contact:

Acquisitions Department
Peter Lang Publishing
275 Seventh Avenue 28th Floor
New York, NY 10001

To order other books in this series, please contact our Customer Service Department:

(800) 770-LANG (within the U.S.)
(212) 647-7706 (outside the U.S.)
(212) 647-7707 FAX

or browse online by series:

WWW.PETERLANGUSA.COM

FSC
www.fsc.org
MIX
Papier | Fördert gute Waldnutzung
FSC® C083411

Kolibri 360 GmbH
Ferdinand-Jühlke-Straße 7
99095 Erfurt, Deutschland
produktsicherheit@kolibri360.de

Druck:
CPI Druckdienstleistungen GmbH
im Auftrag der
Kolibri 360 GmbH
Ferdinand-Jühlke-Str. 7
99095 Erfurt